Lecture Notes in Mathematics 1693

Springer
Berlin
Heidelberg
New York
Barcelona
Budapest
Hong Kong
London
Milan
Paris
Singapore
Tokyo

Stephen Simons

Minimax and Monotonicity

Author

Stephen Simons
Department of Mathematics
University of California
Santa Barbara
CA 93106-3080, USA
simons@math.ucsb.edu

Cataloging-in-Publication Data applied for

Die Deutsche Bibliothek - CIP-Einheitsaufnahme

Simons, Stephen:
Minimax and monotonicity / Stephen Simons. - Berlin ; Heidelberg ; New York ; London ; Paris ; Tokyo ; Hong Kong ; Barcelona ; Budapest : Springer, 1998
(Lecture notes in mathematics ; 1693)
ISBN 3-540-64755-4

Mathematics Subject Classification (1991): 47H05, 47H04, 46B10, 49J35, 47N10

ISSN 0075-8434
ISBN 3-540-64755-4 Springer-Verlag Berlin Heidelberg New York

Printed in Germany

Typesetting: Camera-ready $\mathrm{T_EX}$ output by the author
SPIN: 10649937 46/3143-543210 - Printed on acid-free paper

For Jacqueline

Preface

These notes had their genesis in three hours of lectures that were given in a "School" on minimax theorems that was held in Erice, Sicily in September — October, 1996. This was followed by an expanded version in five hours of lectures at the "Spring School" on Banach spaces in Paseky in the Czech Republic in April, 1997 which was followed, in turn, by an even more expanded version in ten hours of lectures at the University of Toulouse, France in May – June, 1997.

The lectures were initially conceived as three isolated applications of minimax theorems to the theory of monotone multifunctions. With each successive iteration, the emphasis gradually shifted to an examination of the "minimax technique", a method for proving the existence of continuous linear functionals on a Banach space, and to the relationship between this technique and monotone multifunctions. To this was finally added an attempt to collect together the results that have been proved on monotone and maximal monotone multifunctions on Banach spaces in recent years, and organize them into a coherent theory.

I would like to thank many people for their help and encouragement during the various stages of this project. I would first like to thank Biagio Ricceri for inviting me to Erice, Jaroslav Lukes, Jiri Kottas and Vaclav Zizler for inviting me to Paseky, and Jean–Baptiste Hiriart–Urruty for inviting me to Toulouse. I appreciate not only their excellent qualities as hosts, but also their grace and patience as audiences. Thanks are also due to Jonathan Borwein, Simon Fitzpatrick, Simeon Reich and Constantin Zalinescu for reading preliminary versions (or precursors) of these notes, and making many insightful comments and suggestions. I am especially grateful to Heinz Bauschke for reading a semi–final version of these notes from beginning to end, finding an amazing number of errors and ambiguities, and also for providing a number of excellent mathematical ideas. Last, but certainly not least, I would like to express my debt to Robert Phelps for his help and guidance all through this project. I appreciate his dogged insistence that I should try and make these notes as readable as possible. I would also like to acknowledge that his "Prague and Paseky" notes (which have been available electronically for several years) have been a source of inspiration.

Of course, despite all the excellent efforts of the people mentioned above, these notes doubtless still contain errors and ambiguities, and also doubtless have other stylistic shortcomings. At any rate, I hope that there are not too many of these. Those that do exist are entirely my fault.

Stephen Simons
May 28, 1998
Santa Barbara
California

Table of Contents

Chapter IV. General monotone multifunctions

Chapter V. The sum problem for reflexive spaces

Chapter VI. Special maximal monotone multifunctions

Chapter VII. Subdifferentials

Introduction

The primary purpose of these notes is to collect together in one place a number of results that have been proved in recent years about monotone multifunctions on a (possibly nonreflexive) Banach space.

Many of these results involve finding an element of a dual space satisfying certain properties. Ultimately, the solution of such problems relies on the Hahn–Banach theorem. However, it is frequently not easy to apply the Hahn–Banach theorem directly, since this involves the manipulation of sublinear functionals defined by extremely cumbersome formulae.

It turns out that the most convenient way of applying the Hahn–Banach theorem for the problems that we will be considering is through the vehicle of a *minimax theorem*. In fact, one can formulate this procedure into what we will call the "minimax technique". The idea behind this is to use the minimax theorem, the one-dimensional Hahn–Banach theorem and the Banach–Alaoglu theorem to transform problems on the existence of elements of the dual space into problems on the existence of a certain real constant. We will use this technique many times. However, we must emphasize that the minimax theorem as we use it is essentially a repackaging of the Hahn–Banach theorem.

So our secondary purpose is to describe the use of the minimax theorem as a functional analytic tool. The minimax technique does have one great virtue: in many cases one can work backwards from a conjecture, find whether it is reasonable and, if it is, obtain a proof. Even when a result is already known, the proof of it obtained from the minimax technique is usually as simple, if not simpler, than that obtained from more "usual" techniques. Furthermore, isofar as a study of monotonicity is concerned, we feel that the additional work associated with an understanding of the minimax technique is more than justified by the additional insight that this approach permits.

The first result on monotone multifunctions that we will consider will be Rockafellar's "surjectivity" characterization of those monotone multifunctions on a reflexive space that are maximal. In order to do this, we will introduce the "big convexification" of a multifunction, a concept that will be extremely useful to us all through these notes.

The analysis of convex lower semicontinuous functions is, by and large, simpler than the analysis of multifunctions. It is with this in mind that we show how to associate convex lower semicontinuous functions with any multifunction. It turns out that these functions capture enough of the structure of the multifunction for us to obtain simple proofs of a number of the known results about monotone multifunctions on (possibly nonreflexive) Banach spaces. For instance, we give simple proofs that *any nontrivial monotone multifunction is locally bounded at any absorbing point of its domain* and also, using the minimax technique, that *the interior of the domain of any maximal monotone multifunction is convex*. In fact, we can give a *precise description* of this interior in terms of the "essential domain" of one of the associated convex functions mentioned above. We can also prove the stronger result that *any point surrounded by the domain of a maximal monotone multifunction is necessarily an interior point of that domain*.

Since we do not assume that the reader has any prior knowledge of convex analysis, we take time off to establish the results that we will need in this area, some of which depend ultimately on Baire's theorem. In particular, we introduce the "dom–dom lemma", a generalization of the classical open mapping theorem.

Using the dom–dom lemma and the minimax technique, we discuss one of the most fascinating questions about monotone multifunctions: *when is the sum of maximal monotone multifunctions on a reflexive Banach space maximal monotone*. We give a proof of Rockafellar's original result, and unify a number of other results that have been proved in this direction recently. In fact, we will give a necessary and sufficient for the sum of maximal monotone multifunctions on a reflexive Banach space to be maximal monotone.

We will discuss some of the subclasses of the class of maximal monotone multifunctions that have been introduced over the years. The oldest of them, the maximal monotone multifunctions of "type (D)", dates back to 1971, while those that are of "type (FP)", "type (FPV)", "type (NI)", "type (ANA)" and those that are "strongly maximal monotone" are much more recent. We give the definitions of these subclasses, prove the results known about them, and also discuss a number of related open problems.

Rockafellar also proved that *the subdifferential of a somewhere finite convex lower semicontinuous function is maximal monotone*. We give a proof of this result that relies on Ekeland's variational principle and the minimax technique, and then consider generalizations of this result related to the subclasses of multifunctions discussed above.

Using the minimax technique many times, we also consider (possibly unbounded) positive linear operators from a Banach space into its dual. We give a criterion for such an operator to be maximal monotone, and discuss the relationship between such operators and the subclasses of multifunctions discussed above.

In Chapter I, we give the results from functional analysis on which these notes will be based. Starting from the Hahn-Banach theorem for sublinear functionals, we deduce in Theorem 1.1(b) the version of it due to Mazur and Orlicz, an extremely useful result that is not nearly as well known as it deserves to be. We prove the minimax theorem in Theorem 3.1. In Section 4, we introduce the results from Banach space theory that we shall need. The most important of these is the Banach–Alaoglu theorem, Theorem 4.1. Thus Sections 1–4 contain everything we need in order to use the minimax technique. We also define reflexivity in Section 4. In general terms, one of the big problems about monotone multifunctions is finding when results that are known in the reflexive case can be extended to the nonreflexive case. This is why we have included Section 5. In this section, we show that there are "natural boundaries" to minimax theorems in the sense that if a nonempty bounded closed convex set in a Banach space has the property that minimax theorems always hold on it (in a reasonable sense) then the set is necessarily weakly compact. If this set is a closed ball then, from Theorem 4.3, the Banach space is necessarily reflexive.

Section 6 is about the minimax technique. The results in this section will not be used until Section 33. We start off Section 6 by reproving two results that most readers will already be familiar with, the extension form of the Hahn–Banach theorem and the "point – closed convex set" separation theorem in a Banach space. The second of these gives our first concrete example of how the minimax technique can be used to transform a problem on the existence of a linear functional into a problem on the existence of a real constant. The next two examples of the minimax technique will not be familiar to readers who do not have a background in convex analysis. In the second of these, Example 6.2, we give a necessary *and sufficient* form of the Fenchel duality theorem. The minimax technique approach avoids the aggravating problem of the "vertical hyperplane" that so destroys the elegance of the usual approach through the Eidelheit separation theorem. Section 6 closes with some additional remarks on Fenchel conjugates of convex functions.

So now we have transformed our problem on the existence of a linear functional into a problem on the existence of a real constant, how do we find the real constant? What leaps to mind is, of course, Baire's theorem. This is frequently the case — in fact, Chapter III will discuss some new results on convex functions that have been spawned by this problem. However, there is a significant case where this constant can be produced without Baire's theorem. This will be dealt with in Section 7. The main result here is the "fg–theorem" Theorem 7.2. The fg–theorem is unusual in that it uses the minimax theorem twice, first to produce a real constant and then, using this constant as a bound, to apply the minimax technique as described above. Though monotonicity is not mentioned in it, the fg–theorem is, in fact, an abstraction of results on monotonicity that appeared in our paper [54]. Thus Section 7 is a bridge between functional analysis and monotonicity.

In Chapter II, we give the definitions and develop the machinery that we will use to prove one direction of Rockafellar's surjectivity theorem (see below). We define multifunctions, monotonicity and maximality formally in Section 8. For some problems, it is convenient to think of a multifunction from E into 2^{E^*} (for the rest of this introduction, E is a nonzero real Banach space) as a subset of $E \times E^*$. This is pursued in Section 9, where we introduce a *big convexification* of any nonempty subset of $E \times E^*$ and the associated linear operators p, q and r. We also prove the "pqr–lemma", in which monotone subsets of $E \times E^*$ are characterized in terms of their big convexifications. We will use the pqr–lemma many times in these notes.

Section 10 is about reflexive spaces. In Lemma 10.1, we apply the minimax technique to obtain an equivalence valid for any nonempty subset of $E \times E^*$. We use this in Theorem 10.6 to prove that *if M is a monotone subset of $E \times E^*$ then*

$$M \text{ is maximal monotone} \iff M + G(-J) = E \times E^*,$$

where J is the duality map, and we deduce in Theorem 10.7 that

$$S\colon E \mapsto 2^{E^*} \text{ is maximal monotone} \quad \Longrightarrow \quad R(S+J) = E^*.$$

This is one direction of Rockafellar's "surjectivity theorem", except that our results hold for any reflexive space, not only ones where the norms of E and E^* are strictly convex. In the final section of this chapter, Section 11, we use the minimax technique to prove the following useful result:

$$S\colon E \mapsto 2^{E^*} \text{ maximal monotone and } R(S) \text{ bounded} \quad \Longrightarrow \quad D(S) = E.$$

The flowchart below should serve to show which sections are needed for an understanding of Chapter II.

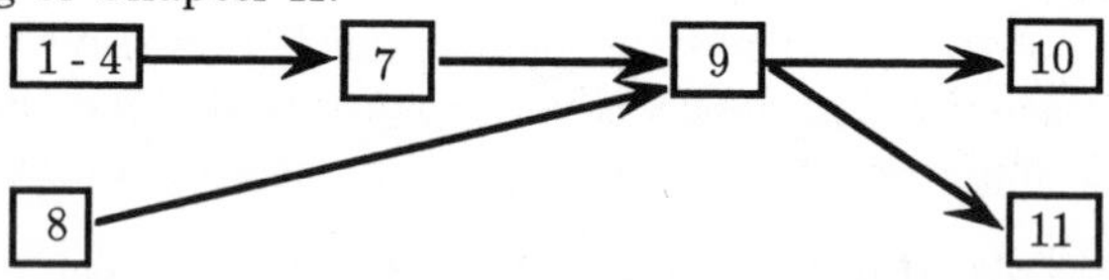

In Chapter III, we leave our discussion of monotonicity temporarily, and turn our attention to convex analysis. We collect together in Sections 12 and 13 various results on convex functions that depend ultimately on Baire's theorem. The "dom lemma", Lemma 12.2, is a generalization to convex functions of the classical uniform boundedness (Banach Steinhaus) theorem (see Remark 12.4) and the "dom–dom lemma", Lemma 13.1 is a generalization to convex functions of the classical open mapping theorem (see Remark 13.3). Both of these results will be applied later on to obtain bounds that can be used for the minimax technique. (We should, however, remind the reader that the bound β of the fg–theorem, Theorem 7.2 was found *without* the use of Baire's theorem.) We can think of the dom lemma and the dom–dom lemma as "quantitative" results, since their main purpose is to provide numerical bounds. Associated with them are two "qualitative" results, the "dom corollary", Corollary 12.3, and the "dom–dom corollary", Corollary 13.2, from which the numerics have been removed. These results will also be of use to us later on. We give in Remark 13.4 a brief discussion of convex Borel sets and functions. In the final section of this chapter, Section 14, we show how the dom–dom lemma leads to the Attouch–Brézis version of the Fenchel duality theorem, which we state formally as Theorem 14.2. Theorem 14.2 will not be used in our later work on monotonicity, however we thought that it would be appropriate to show how the dom–dom lemma enables us to complete Example 6.2. The next flowchart shows which sections are needed for an understanding of Chapter III.

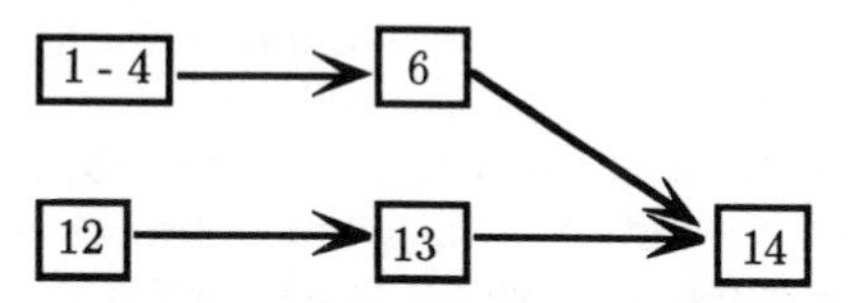

In Chapter IV, we end our digression into convex analysis, and return to our analysis of multifunctions. As we have already explained, we can get considerable insight into the properties of multifunctions by considering associated convex functions (and then applying the results of the Chapter III to them). In Section 15, we define two such convex functions ψ_S and χ_S associated with every nontrivial multifunction $S\colon E \mapsto 2^{E^*}$. The formula for χ_S is more complicated than that for ψ_S. However, the set indexing χ_S is convex, which will enable us to use the minimax technique. We continue this analysis in Section 16 by showing how ψ_S and χ_S interact with closed convex sets and closed subspaces. The result is particularly simple when S is maximal monotone and $D(S)$ is closed and convex — we shall show in Theorem 16.2 that, in this case,

$$\operatorname{dom} \chi_S = \operatorname{dom} \psi_S = D(S).$$

(The proof, however, is not so simple: it uses either Rockafellar's result on the maximal monotonicity of subdifferentials, or the Bishop–Phelps theorem.) The remainder of Section 16 is devoted to proving some results that we shall need for our analysis of constraint qualifications for pairs of maximal monotone multifunctions. In particular, in Theorem 16.10, we deal with the problem of "restricting" a maximal monotone multifunction to a closed subspace.

In Section 17, using the dom lemma, we establish results that we will use in Sections 18 and 20, and we also prove a local boundedness theorem for any (not necessarily monotone) multifunction on a Banach space. Specifically, we prove in Theorem 17.3 that a *nontrivial multifunction is locally bounded at each point surrounded by* $\operatorname{dom} \psi_S$. This extends the results known for monotone multifunctions. In Section 18, we use the minimax technique to prove two main results, the "six set theorem", Theorem 18.3, and the "nine set theorem", Theorem 18.4. For both of these, we assume that S is maximal monotone. In the six set theorem, we prove that the six sets $\operatorname{int} D(S)$, $\operatorname{int}(\operatorname{co} D(S))$, $\operatorname{int}(\operatorname{dom}\chi_S)$, $\operatorname{sur} D(S)$, $\operatorname{sur}(\operatorname{co} D(S))$ and $\operatorname{sur}(\operatorname{dom}\chi_S)$ coincide, and in the nine set theorem, we prove that, if $\operatorname{sur}(\operatorname{dom}\chi_S) \neq \emptyset$, then the nine sets $\overline{D(S)}$, $\overline{\operatorname{co} D(S)}$, $\overline{\operatorname{dom}\chi_S}$, $\overline{\operatorname{int} D(S)}$, $\overline{\operatorname{int}(\operatorname{co} D(S))}$, $\overline{\operatorname{int}(\operatorname{dom}\chi_S)}$, $\overline{\operatorname{sur} D(S)}$, $\overline{\operatorname{sur}(\operatorname{co} D(S))}$ and $\overline{\operatorname{sur}(\operatorname{dom}\chi_S)}$ coincide. ("Sur" is defined in Section 12.) These results strengthen results of Rockafellar that $\operatorname{int} D(S)$ is convex and that, if $\operatorname{int}(\operatorname{co} D(S)) \neq \emptyset$ then $\overline{D(S)}$ is convex, and also settle in the affirmative an open problem as to whether an absorbing point of $D(S)$ is necessarily an interior point. We do not know if the results analogous to the six set theorem and the nine set theorem hold with "χ_S" replaced by "ψ_S". We end Section 18 by specializing to the reflexive case, in which the answer to the above question is in the affirmative. In Section 19, we introduce a convex function ξ_S "dual" to ψ_S and show the connection between this function and Brézis–Haraux appproximation, which is concerned with finding conditions under which $R(S_1 + S_2)$ is "approximately equal" to $R(S_1) + R(S_2)$ in the sense that

$$\left.\begin{aligned} \overline{R(S_1+S_2)} &= \overline{R(S_1)+R(S_2)} \\ &\text{and} \\ \operatorname{int}\left[R(S_1+S_2)\right] &= \operatorname{int}\left[R(S_1)+R(S_2)\right]. \end{aligned}\right\} \tag{19.0.1}$$

Brézis–Haraux give two conditions ((19.0.2) and (19.0.3)) which imply (19.0.1). We show that each of these imply that

$$R(S_1) + R(S_2) \subset \operatorname{dom}\xi_{S_1+S_2}, \tag{19.3.1}$$

and that (19.3.1) in turn implies (19.0.1). There is a third condition, due to Pazy, which implies (19.3.1). We do not know if Pazy's condition implies (19.3.1). (See Problem 19.6.) The next flowchart shows which sections are needed for an understanding of Chapter IV.

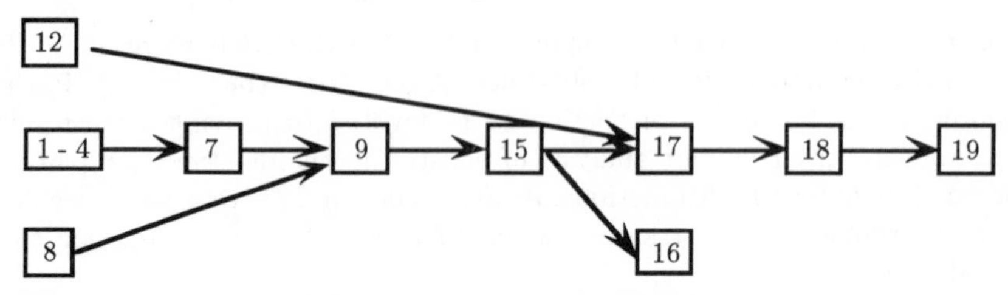

Chapter V is motivated by the following result of Rockafellar and recent generalizations of it. *If E is reflexive, S_1 and S_2 are maximal monotone and*

$$D(S_1) \cap \operatorname{int} D(S_2) \neq \emptyset \tag{20.0.1}$$

then $S_1 + S_2$ is maximal monotone. Indeed, we give a proof of this result in Theorem 20.5. Apart from this, Section 20 is devoted to setting up the machinery that we shall need for the more general results that appear later on in the chapter. Much of the discussion centers on the "γ–condition" (20.1.1). We shall explain in the text why this ugly condition is, in fact, forced on us by the nature of our problem. We then use the minimax technique to prove in Lemma 20.1 the equivalence of the γ–condition with condition (20.1.2), which says that there exist x_1^*, $x_2^* \in E^*$ and $z \in E$ such that, for all $(s_1, s_1^*) \in G(S_1)$ and $(s_2, s_2^*) \in G(S_2)$,

$$2\langle s_1 - z, s_1^* - x_1^*\rangle + 2\langle s_2 - z, s_2^* - x_2^*\rangle \geq \|z\|^2 + \|x_1^* + x_2^*\|^2 + 2\langle z, x_1^* + x_2^*\rangle.$$

An argument due originally to Minty and Browder enables us to deduce in Lemma 20.2 that there exists $(z, z^*) \in G(S_1 + S_2)$ such that

$$\|z\|^2 + \|z^*\|^2 + 2\langle z, z^*\rangle = 0.$$

Using an extension of the local boundedness theorem established in Lemma 17.2(a), we then prove in Lemma 20.3 that Rockafellar's condition, (20.0.1), implies the γ–condition, and so all the results above hold. Our final step towards Theorem 20.5 is Lemma 20.4, in which we boostrap by translating first in E and then in E^*. Theorem 20.5 itself is proved by using the criterion for maximality in Theorem 10.3. The next flowchart shows which sections are needed for an understanding of Section 20.

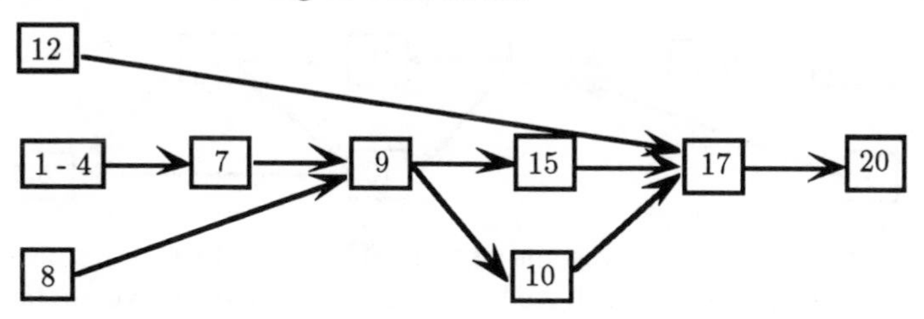

Section 21 is about the "dom–dom constraint qualification",

$$\operatorname{dom} \chi_{S_1} - \operatorname{dom} \chi_{S_2} \quad \text{is absorbing.} \tag{21.0.1}$$

Indeed, we shall prove in Theorem 21.3 that the maximal monotonicity of $S_1 + S_2$ remains true even if (20.0.1) is weakened to (21.0.1). This result is established using exactly the same steps as in Section 20, except that Lemma 21.1 is much harder than the result to which it corresponds, Lemma 20.3. (Lemma 21.1 uses the dom–dom lemma, Lemma 13.1 rather than the dom lemma, Lemma 12.2.) Actually, our analysis contains a necessary and sufficient condition (still, for reflexive spaces) for the sum of maximal monotone multifunctions to be maximal monotone. We have set this out in Theorem 21.4, but we suspect that it may be too complicated to be of any practical use.

In Section 22, we prove that the six set theorem and the nine set theorem established in Section 18 for $D(S)$ (S maximal monotone on a general Banach space) have analogs for $D(S_1) - D(S_2)$ (S_1 and S_2 maximal monotone on a reflexive Banach space). We deduce from this in Section 23 the equivalence of several constraint qualifications that have been proposed recently. The next flowchart shows which sections are needed for an understanding of Sections 21—23.

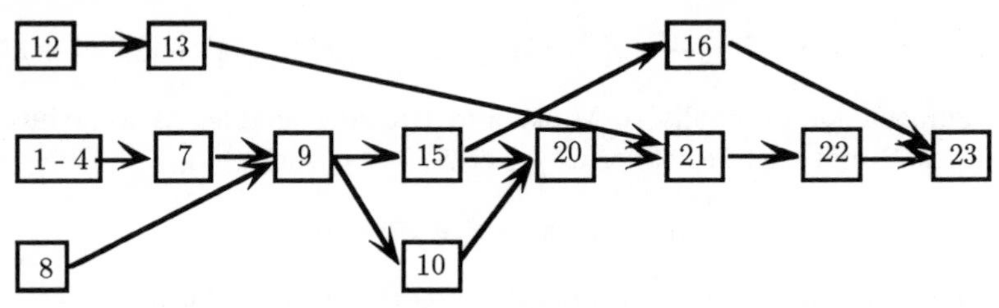

We close the chapter by showing in Section 24 how the techniques of these notes can be used to establish the Brézis–Crandall–Pazy result on the maximal monotonicity of the sum. The final flowchart for Chapter V shows which sections are needed for an understanding of Section 24. (The only part of Section 20 that is needed is Lemma 20.2, which does not depend on Baire's theorem.)

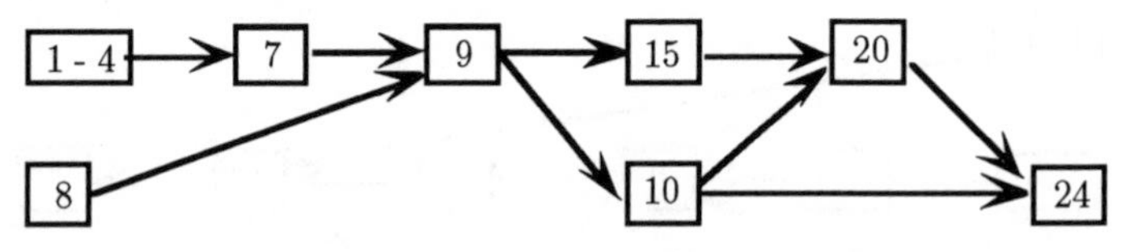

In Chapter VI, we discuss some of the subclasses of the class of maximal monotone multifunctions that have been introduced over the years. In Section 25, we define and give the basic properties of those that are of "type (D)", "type (FP)", "type (FPV)", "type (NI)", "type (ANA)", and those that are "strongly maximal monotone".

Now suppose that E is nonreflexive and S is maximal monotone. It has been noted in Problem 18.9 that we do not know if $\overline{D(S)}$ is necessarily convex, and it is also noted in the remarks preceding Definition 25.4 that we do not know if S is necessarily of type (FPV). The connection between these observations is clarified in Section 26. Indeed, we prove in Theorem 26.3 that if $\overline{D(S)} \neq \overline{\operatorname{dom}\psi_S}$ (in particular, if $\overline{D(S)}$ is not convex) then S is not of type (FPV), and in Theorem 26.1 that if S is not of type (FPV) then we have a negative answer to the long–standing question whether Rockafellar's sum theorem is true in nonreflexive spaces. In Section 27, we return to our consideration of the function ξ_S introduced in Section 19, and show that in the three cases where it has been proved that $\overline{R(S)}$ is convex one can, in fact, prove that $\overline{R(S)} = \overline{\operatorname{dom}\xi_S}$. Some of the results of Section 27 are "dual" to those of Section 26, while others seem to rely on totally different principles.

The next flowchart shows which sections are needed for an understanding of Chapter VI. (The only part of Section 19 needed in Section 27 is the definition of ξ_S.)

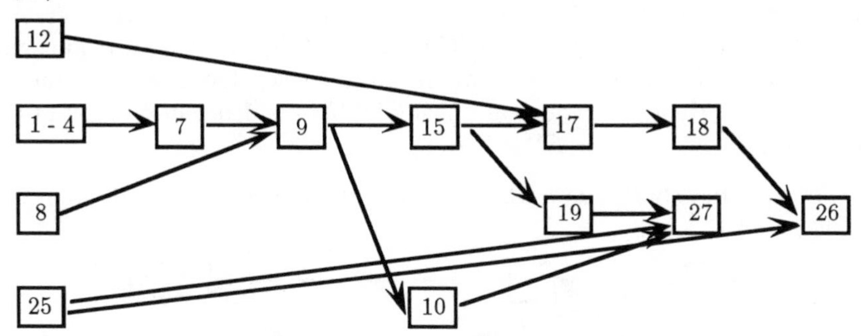

In Chapter VII, we consider the properties of subdifferentials. There are various ways of approaching this topic. In the end, the easiest seems to be through the formula for the subdifferential of the sum of two convex functions, which we consider in Section 28. The main work for this is actually in Lemma 28.1 (a form of the Fenchel duality theorem in which we do not explicitly mention conjugates). We establish this using the minimax technique and a "scaling" argument, while the formula for the subdifferential of the sum is obtained by a simple bootstrapping procedure in Theorem 28.2. In Section 29, we use the minimax technique and Ekeland's variational principle to obtain a version of the Brøndsted–Rockafellar theorem, which we combine with the formula for the subdifferential of a sum to obtain our main result on the existence of subgradients, Theorem 29.4. With one significant exception, all the later results on subdifferentials depend on this theorem. In Corollary 29.5, we obtain (modulo some simple bootstrapping) Rockafellar's classical result that subdifferentials are maximal monotone. The main result in Section 30 is Theorem 30.3, in which we establish that subdifferentials are of type (FP), and the main result in Section 31 is Theorem 31.3, in which we establish that subdifferentials are of type (FPV). In Section 32, we prove that sudifferentials are strongly maximal monotone. This is the case for which Theorem 29.4 does not seem to be adequate. In order to handle this, we prove in Lemma 32.1 a generalization of Corollary 29.5 in which $\{0\}$ is replaced by any nonempty $w(E, E^*)$–compact convex subset of E.

In Section 35, we shall prove that subdifferentials are maximal monotone of type (D). In fact, we shall define a slightly stronger property ("type (DS)") and prove in Theorem 35.3 that subdifferentials enjoy this stronger property. This implies, in particular, that subdifferentials are of "dense type" in the sense of Gossez. Gossez's analysis relies on the theory of locally convex spaces. We use instead some properties of the biconjugates of convex functions. The preliminary work for Theorem 35.3 is contained in Sections 33 and 34. The main result of Section 33 is Theorem 33.3, in which we establish the formula for the biconjugate of the pointwise maximum of a finite number of functions.

What is curious is that we can establish this result in a situation in which we do not have a simple explicit formula for the *conjugate* of the pointwise maximum. In Section 34, we will define a new topology $\mathcal{T}_{\mathcal{CLB}}(E^{**})$ on E^{**} which lies between the weak* topology $w(E^{**},E^*)$ and the norm topology $\mathcal{T}_{\|\ \|}(E^{**})$. The main result here is Theorem 34.7, in which we use Theorem 33.3 to show that if f is a somewhere finite convex lower semicontinuous function on E then the canonical image of $G(\partial f)$ is dense in $G(\partial f^*)$ in the topology $\mathcal{T}_{\|\ \|}(E^*) \times \mathcal{T}_{\mathcal{CLB}}(E^{**})$.

The next flowchart shows which sections are needed for an understanding of Chapter VII.

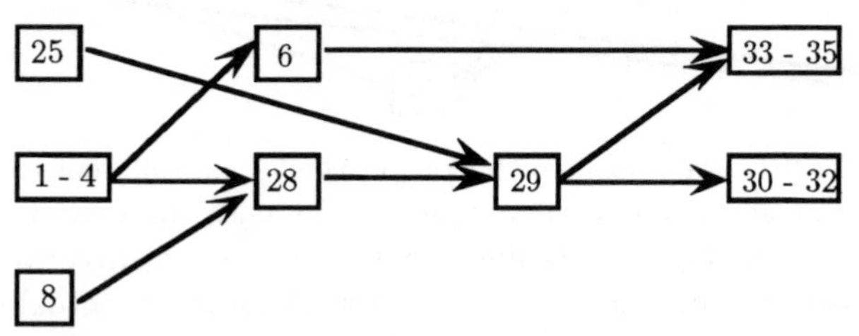

However, readers already familiar with the formula for the subdifferential of a sum, the Brøndsted–Rockafellar theorem and the elementary properties of the Fenchel conjugate should be able to read most of Chapter VII without needing to refer to the previous chapters of these notes.

We give a brief discussion of some of the properties of (possibly unbounded) positive *linear* operators in Chapter VIII. Theorem 36.2 contains a necessary and sufficient condition for a positive linear operator to be maximal monotone. Theorem 37.1 gives a sufficient condition for the sum of two maximal monotone linear operators to be maximal monotone. We shall see that, for linear operators, Rockafellar's original sum theorem is true *even without the assumption of reflexivity*. In Section 38, we consider which positive linear operators belong to various subclasses of the family of maximal monotone multifunctions introduced in Section 25: we use the minimax technique to prove in Theorem 38.2 that every maximal monotone linear operator is also of type (FPV), in Theorem 38.3 that every linear operator that is maximal monotone of type (NI) is also of type (FP), in Theorem 38.5 that every maximal monotone linear operator is also strongly maximal monotone, and in Theorem 38.6 that every continuous positive linear operator is of type (ANA). The next flowchart shows which sections are needed for an understanding of Chapter VIII.

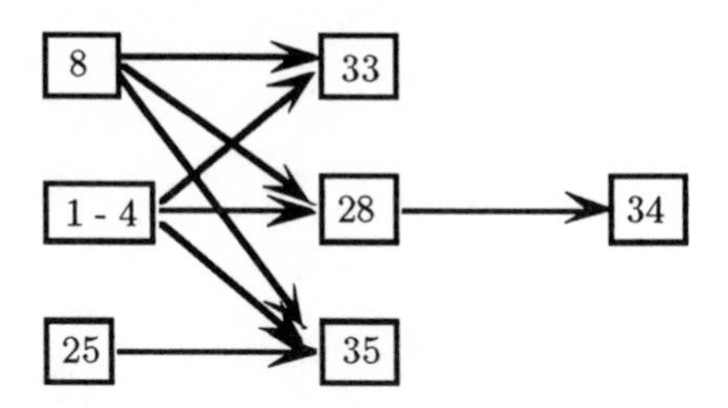

In Chapter IX, we shall give some cases in which Rockafellar's sum theorem (see the discussion of Chapter V above) is true in the nonreflexive case.
• We shall prove in Theorem 40.4 the result of Heisler that *if* $S_1\colon E \mapsto 2^{E^*}$ *and* $S_2\colon E \mapsto 2^{E^*}$ *are maximal monotone and* $D(S_1) = D(S_2) = E$ *then* $S_1 + S_2$ *is maximal monotone.*
• We shall prove in Theorem 41.2 that *if* C *is a nonempty closed convex subset of* E, $S\colon E \mapsto 2^{E^*}$ *is maximal monotone and* $\operatorname{int} D(S) \supset C$ *then* $S+N_C$ *is maximal monotone.* (Here N_C is the *normality multifunction* associated with the set C, defined in Section 8.)
• We shall prove in Theorem 41.6 that *if* C *is a closed convex subset of* E, $D(S)$ *is a subspace of* E, $S\colon D(S) \mapsto E^*$ *is linear and maximal monotone and* $D(S) \cap \operatorname{int} C \neq \emptyset$ *then* $S + N_C$ *is maximal monotone.*
• We shall prove in Theorems 42.1 and 42.2 the two results of Bauschke that *if* $S\colon E \mapsto 2^{E^*}$ *is maximal monotone and* $T\colon E \mapsto E^*$ *is skew and linear then* $S+T$ *is maximal monotone* and *if* f *is a somewhere finite convex lower semicontinuous function on* E *and* $T\colon E \mapsto E^*$ *is positive and linear then* $\partial f + T$ *is maximal monotone.*

The last two flowcharts show which sections are needed for an understanding of Chapter IX.

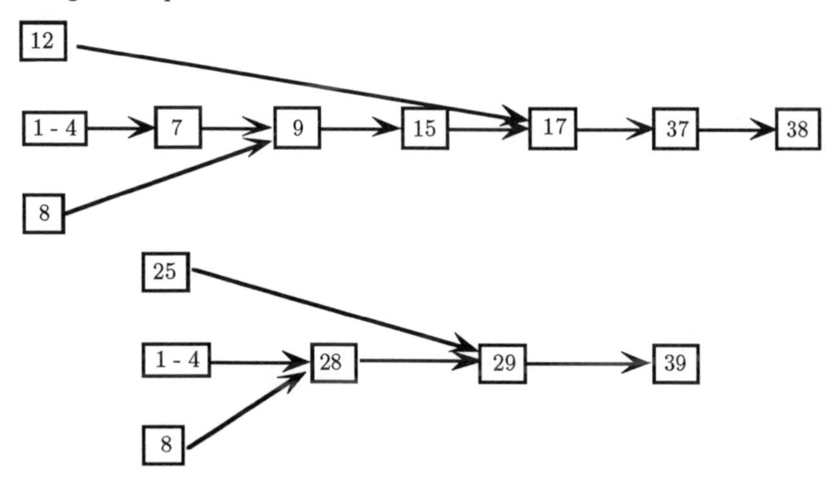

In the final chapter, Chapter X, we collect together some of the open problems that have appeared in the body of the text.

I. Functional analytic preliminaries

1. The Hahn–Banach and Mazur–Orlicz theorems

We take as our starting point the "sublinear form" of the Hahn–Banach theorem, and the Mazur–Orlicz theorem. Let E be a nonzero real vector space. For the moment we do not need any additional structure for E. A *sublinear functional on* E is a map $S\colon E \mapsto \mathbb{R}$ such that

$$x,\ y \in E \quad \Longrightarrow \quad S(x+y) \leq S(x) + S(y) \tag{1.0.1}$$

and

$$x \in E \text{ and } \lambda > 0 \quad \Longrightarrow \quad S(\lambda x) = \lambda S(x). \tag{1.0.2}$$

It follows from (1.0.2) that $S(0) = S(2 \times 0) = 2 \times S(0)$, hence $S(0) = 0$. Consequently, (1.0.2) can be strengthened to:

$$x \in E \text{ and } \lambda \geq 0 \quad \Longrightarrow \quad S(\lambda x) = \lambda S(x).$$

Note that a norm or a seminorm is an example of a sublinear functional. So also are linear functionals. Since nontrivial linear functionals are never positive, and norms and seminorms are, by definition, always positive, there are plenty of examples of sublinear functionals that are not norms or seminorms. See the paper [31] by König for some of the subtler properties of sublinear functionals.

The Hahn–Banach theorem, Theorem 1.1(a), is undoubtedly the most important result in functional analysis. Theorem 1.1(b) is the generalization of the Hahn–Banach theorem due to Mazur and Orlicz. The Mazur–Orlicz theorem is not nearly as well known as it deserves to be — we refer the reader to the paper [30] by König for a number of applications of it to other fields of analysis.

Theorem 1.1. *Let S be a sublinear functional on E. Then:*
(a) There exists a linear functional L on E such that

$$L \leq S \text{ on } E.$$

(b) In addition, let C be a nonempty convex subset of E. Then there exists a linear functional L on E such that

$$L \leq S \text{ on } E \quad \text{and} \quad \inf_C L = \inf_C S.$$

Proof. (a) See Kelly–Namioka, [28], 3.4, p. 21 for a proof using cones, Rudin, [46], Theorem 3.2, p. 56–57 for a proof using an extension by subspaces argument, and [47] for a proof using an ordering on sublinear functionals. This last reference contains an integrated treatment of the Hahn–Banach theorem, the Mazur–Orlicz theorem and the Choquet–Bishop–de Leeuw results on the existence and uniqueness of representing measures.

(b) Let

$$\alpha := \inf_C S.$$

If $\alpha = -\infty$, the result is immediate from (a) (take any linear functional L on E such that $L \leq S$ on E). So we can suppose that $\alpha \in \mathbb{R}$. Define $T\colon E \mapsto \mathbb{R} \cup \{-\infty\}$ by

$$T(x) := \inf_{y \in C,\ \lambda > 0} [S(x + \lambda y) - \lambda\alpha]. \tag{1.1.1}$$

If $x \in E$, $y \in C$ and $\lambda > 0$ then $\alpha \leq S(y)$ hence

$$\begin{aligned} S(x + \lambda y) - \lambda\alpha &\geq S(x + \lambda y) - \lambda S(y) \\ &= S(x + \lambda y) - S(\lambda y) \geq -S(-x) > -\infty. \end{aligned}$$

Taking the infimum over $y \in C$ and $\lambda > 0$, $T(x) \geq -S(-x) > -\infty$. Thus $T\colon E \mapsto \mathbb{R}$. It is now easy to check that T is a sublinear functional (exercise!). From the Hahn–Banach Theorem, there exists a linear functional L on E such that

$$L \leq T \text{ on } E.$$

Letting $\lambda \to 0$ in (1.1.1), $T \leq S$ on E (exercise!). Thus $L \leq S$ on E, as required. Let $x \in C$. Then

$$-L(x) = L(-x) \leq T(-x) \leq S(-x + 1x) - 1\alpha = -\alpha.$$

Hence $L(x) \geq \alpha$. Taking the infimum over $x \in C$, $\inf_C L \geq \alpha = \inf_C S$. On the other hand, since $L \leq S$ on E, $\inf_C L \leq \inf_C S$. ∎

Our next result, the "one–dimensional Hahn–Banach theorem" (which can also be deduced from the "extension form" of the Hahn–Banach theorem, see Rudin, [46], Theorem 3.2, p. 56–57 (exercise!)) follows immediately from the Mazur–Orlicz theorem, Theorem 1.1(b) by taking $C := \{x\}$.

Corollary 1.2. *Let S be a sublinear functional on E and $x \in E$. Then there exists a linear functional L on E such that*

$$L \leq S \text{ on } E \quad \text{and} \quad L(x) = S(x).$$

2. Convex, concave and affine functions

Let X be a nonempty convex subset of a vector space, and $f: X \mapsto \mathbb{R}$. We say that f is *convex* if

$$x, y \in X \text{ and } \lambda \in (0,1) \quad \Longrightarrow \quad f(\lambda x + (1-\lambda)y) \leq \lambda f(x) + (1-\lambda) f(y).$$

We say that f is *concave* if

$$x, y \in X \text{ and } \lambda \in (0,1) \quad \Longrightarrow \quad f(\lambda x + (1-\lambda)y) \geq \lambda f(x) + (1-\lambda) f(y).$$

Now let Z also be a convex subset of a vector space and $f: X \mapsto Z$. We say that f is *affine* if

$$x, y \in X \text{ and } \lambda \in (0,1) \quad \Longrightarrow \quad f(\lambda x + (1-\lambda)y) = \lambda f(x) + (1-\lambda) f(y).$$

The following extension of the definition of convex function is motivated by constrained optimization — if X is a nonempty subset of a vector space E, $f: X \mapsto \mathbb{R}$ and we are trying to find a minimum of f over X, we can extend the definition of f to be $+\infty$ on $E \setminus X$, and thereby produce a function defined over E. Clearly a minimum of the extended function over E is identical with a minimum of the original function over X. So we frequently assume that our functions are defined on E, but take values in $\mathbb{R} \cup \{\infty\}$. If then $f\colon E \mapsto \mathbb{R} \cup \{\infty\}$, we write

$$\operatorname{dom} f := \{x \in E\colon\ f(x) \in \mathbb{R}\}.$$

We say that f is "convex" if $\operatorname{dom} f$ is convex and the restriction of f to $\operatorname{dom} f$ is convex in the sense already defined. Equivalently, we can say:

$$x, y \in E \text{ and } \lambda \in (0,1) \quad \Longrightarrow \quad f(\lambda x + (1-\lambda)y) \leq \lambda f(x) + (1-\lambda) f(y),$$

provided that we interpret $\infty + \infty$ to be ∞, and $\lambda \times \infty$ to be ∞ for $\lambda > 0$.

One final point of notation: if λ, $\mu \in \mathbb{R}$, we write $\lambda \vee \mu$ for the maximum value of λ and μ, and $\lambda \wedge \mu$ for the minimum value of λ and μ. The result contained in Lemma 2.1(a) can also be deduced from Fan–Glicksberg–Hoffman, [22], Theorem 1, p. 618, after some simple transformations. We note the total absence of topological hypotheses in Lemma 2.1 — this will be important for us later.

Lemma 2.1. *Let X be a nonempty convex subset of a vector space.*
(a) Let $f_1, \ldots, f_m$ be convex real functions on X. Then there exist $\lambda_1, \ldots, \lambda_m \geq 0$ such that $\lambda_1 + \cdots + \lambda_m = 1$ and

$$\inf_X [f_1 \vee \cdots \vee f_m] = \inf_X [\lambda_1 f_1 + \cdots + \lambda_m f_m].$$

(b) Let $g_1, \ldots, g_m$ be concave real functions on X. Then there exist $\lambda_1, \ldots, \lambda_m \geq 0$ such that $\lambda_1 + \cdots + \lambda_m = 1$ and

$$\sup_X [g_1 \wedge \cdots \wedge g_m] = \sup_X [\lambda_1 g_1 + \cdots + \lambda_m g_m].$$

Proof. (a) Let $E := \mathbb{R}^m$. Define $S\colon E \mapsto \mathbb{R}$ by

$$S(\mu_1, \ldots, \mu_m) := \mu_1 \vee \cdots \vee \mu_m.$$

S is sublinear. Let

$$C := \big\{(\mu_1, \ldots, \mu_m)\colon \text{there exists } x \in X \text{ such that,} \\ \text{for all } i = 1, \ldots, m,\ f_i(x) \le \mu_i\big\}.$$

C is a convex subset of E (exercise!). From the Mazur–Orlicz Theorem, there exists a linear functional L on E such that

$$L \le S \text{ on } E \quad \text{and} \quad \inf_C L = \inf_C S.$$

Since L is linear, there exist $\lambda_1, \ldots, \lambda_m \in \mathbb{R}$ such that,

$$\text{for all } (\mu_1, \ldots, \mu_m) \in \mathbb{R}^m, \quad L(\mu_1, \ldots, \mu_m) = \lambda_1 \mu_1 + \cdots + \lambda_m \mu_m.$$

Since $L \le S$ on E, $\lambda_1, \ldots, \lambda_m \ge 0$ and $\lambda_1 + \cdots + \lambda_m = 1$ (exercise!). Finally,

$$\inf_C L = \inf_{x \in X} [\lambda_1 f_1(x) + \cdots + \lambda_m f_m(x)] = \inf_X [\lambda_1 f_1 + \cdots + \lambda_m f_m]$$

and

$$\inf_C S = \inf_{x \in X} [f_1(x) \vee \cdots \vee f_m(x)] = \inf_X [f_1 \vee \cdots \vee f_m].$$

This gives (a), and (b) follows from (a) by taking $f_i := -g_i$. ∎

3. The minimax theorem

Let A, B be nonempty sets, and $h\colon A \times B \mapsto \mathbb{R}$. We shall write $\inf_A \sup_B h$ for $\inf_{a \in A} \sup_{b \in B} h(a,b)$, and use a corresponding shorthand for other similar expressions. It is easily seen that (exercise!) the inequality

$$\sup_B \inf_A h \le \inf_A \sup_B h \tag{3.0.1}$$

is always satisfied. This inequality can be strict, take for instance $A = B = \{0,1\}$ and $h(a,b) = 0$ if $a \ne b$ and $h(a,b) = 1$ if $a = b$. A *minimax theorem* is a theorem that gives conditions under which

$$\sup_B \inf_A h = \inf_A \sup_B h.$$

One final remark on notation: we shall say that h is "convex on A" if, for all $b \in B$, $h(\cdot, b)$ is convex, and use a corresponding shorthand for other similar expressions.

There are many different minimax theorems (see our survey [52]). In Theorem 3.1, we shall use Lemma 2.1 to give a simple proof of a minimax theorem that follows from a result of Fan (see [21]). (See also the paper [29] by König and our paper [48] for simple generalizations of Fan's result.) It is important to note that the set A has no topological structure. It is worth mentioning that Lemma 2.1(a) can be deduced from Theorem 3.1 (exercise!).

Theorem 3.1. *Let A be a nonempty convex subset of a vector space, B be a nonempty convex subset of a vector space and B also be a compact Hausdorff topological space. Let $h\colon A \times B \mapsto \mathbb{R}$ be convex on A, and concave and upper semicontinuous on B. Then*

$$\inf_A \max_B h = \max_B \inf_A h.$$

Proof. We can write "max" instead of "sup" because h is upper semicontinuous on B and B is compact. Let $\alpha := \inf_A \max_B h$. Let $a_1, \ldots, a_m \in A$. Then, from Lemma 2.1(b) with $X := B$ and $g_i := h(a_i, \cdot)$, there exist $\lambda_1, \ldots \lambda_m \geq 0$ such that $\lambda_1 + \cdots + \lambda_m = 1$ and

$$\max_{b\in B}[h(a_1,b) \wedge \cdots \wedge h(a_m,b)] = \max_{b\in B}[\lambda_1 h(a_1,b) + \cdots + \lambda_m h(a_m,b)].$$

Since h is convex on A,

$$\max_{b\in B}[h(a_1,b) \wedge \cdots \wedge h(a_m,b)] \geq \max_{b\in B} h(\lambda_1 a_1 + \cdots + \lambda_m a_m, b) \geq \alpha.$$

Thus

$$\{b \in B\colon\ h(a_1,b) \geq \alpha\} \cap \cdots \cap \{b \in B\colon\ h(a_m,b) \geq \alpha\} \neq \emptyset.$$

Since B is compact Hausdorff, and h is upper semicontinuous on B, from the "finite intersection property" of the closed sets in B,

$$\bigcap_{a\in A} \{b \in B\colon\ h(a,b) \geq \alpha\} \neq \emptyset.$$

Hence

$$\max_{b\in B} \inf_{a\in A} h(a,b) \geq \alpha,$$

that is to say,

$$\max_B \inf_A h \geq \inf_A \max_B h.$$

The result of Theorem 3.1 now follows from (3.0.1). ∎

4. The dual and bidual of a Banach space

From now on, E will be a nonzero Banach space, and E^* will stand for the *dual space* of E, the set of continuous linear functionals on E. If L is a linear functional on E then L is continuous if, and only if,

$$\|L\| := \sup_{x \in E,\ \|x\| \le 1} L(x) < \infty.$$

If $x^* \in E^*$, we will write $\langle x, x^* \rangle$ for the value of x^* at x. The reason for this notation is that we sometimes want to consider the elements of E as functions on E^* and "$\langle x, \cdot \rangle$" is easier to read than "$\cdot(x)$". However, as we shall see later, despite the more symmetric nature of this notation, the situation is not totally symmetric. Anyhow, with this notation,

$$\text{for all } x^* \in E^*, \quad \|x^*\| = \sup_{x \in E,\ \|x\| \le 1} \langle x, x^* \rangle.$$

Note that "$\|\ \|$" does double duty both as the original norm of E, and as the dual norm that we have defined for E^*. However, no confusion will arise from this. Now the norm of E defines a topology on E. However, it is frequently useful to consider the *weak topology*, $w(E, E^*)$, which is defined as the smallest topology on E with respect to which all the functions $\langle \cdot, x^* \rangle$ $(x^* \in E^*)$ are continuous. We now introduce two other important concepts. The first is the *weak* topology*, $w(E^*, E)$, which is defined as the smallest topology on E^* with respect to which all the functions $\langle x, \cdot \rangle$ $(x \in E)$ are continuous. The main result about $w(E^*, E)$ is the celebrated Banach–Alaoglu theorem:

Theorem 4.1. *If $M \ge 0$ then $\{x^* \in E^*\colon \|x^*\| \le M\}$ is $w(E^*, E)$–compact.*

Proof. See Kelly–Namioka, [28], 17.4, p. 155. ∎

In fact there is a stronger result that we shall have occasion to use, namely (exercise!):

Theorem 4.2. *Let $P\colon E \mapsto \mathbb{R}$ be a continuous positive sublinear functional. Then*

$$\{x^* \in E^*\colon x^* \le P \text{ on } E\} \text{ is } w(E^*, E)\text{–compact.}$$

One is tempted to hope that the result "dual" to Theorem 4.1 is true, that is to say: *if $M \ge 0$ then $\{x \in E\colon \|x\| \le M\}$ is $w(E, E^*)$–compact.* We shall see below that this is not true in general.

The second important concept that we need at this time is that of the "bidual", E^{**} of E, which is defined to be the dual of the normed space E^*. Any element of E "gives" an element of E^{**}. More precisely, we can define a linear operator $\widehat{\ }\colon E \mapsto E^{**}$ such that

$$x \in E \text{ and } x^* \in E^* \quad \Longrightarrow \quad \langle x^*, \widehat{x} \rangle = \langle x, x^* \rangle.$$

In some senses, the map $\widehat{\ }$ gives a very tight connection between E and E^{**}. For instance, $\widehat{\ }$ is an "isometry", that is to say, for all $x \in E$, $\|\widehat{x}\| = \|x\|$ (see Holmes, [27], §16.D, p. 122–123). It follows from this and the completeness of E that if B is a norm–closed subset of E then the image $\widehat{B}$ of B is a norm–closed subset of E^{**}. Also, $\widehat{\ }$ is a homeomorphism of E onto $\widehat{E}$ with respect to $w(E, E^*)$ and $w(E^{**}, E^*)$ (exercise!). The tightest connection would be that $\widehat{E} = E^{**}$. We say that E is "reflexive" when this happens. (See [27], §16.F, p. 125–127). Some of the common Banach spaces are reflexive (for instance, Hilbert spaces and the spaces ℓ_p, $(1 < p < \infty)$), and some are not (for instance, ℓ_1, ℓ_∞, c_0 and $C[0,1]$.) The connection with the concepts introduced in the preceding paragraph is contained in the next result:

Theorem 4.3. *E is reflexive if, and only if, $\{x \in E\colon \|x\| \le 1\}$ is $w(E, E^*)$–compact.*

Proof. See [27], §16.F, p. 126–127. ▌

We now show the connection between the Mazur–Orlicz theorem and another well known result in functional analysis, one of the "separation theorems" (see Rudin, [46], Theorem 3.4, p. 58–59 for a more general result).

Theorem 4.4. *Let D be a nonempty convex subset of E and $x \in E \setminus \overline{D}$. Then there exists $z^* \in E^*$ such that $\sup_D z^* < \langle x, z^* \rangle$.*

Proof. From the Mazur–Orlicz theorem, Theorem 1.1(b), with $S := \|\ \|$ and $C := x - D$, there exists a linear function L on E such that

$$L \le \|\ \| \text{ on } E \quad \text{and} \quad \inf_C L = \inf_C \|\ \|.$$

Since $L \le \|\ \|$ on E, $L \in E^*$ (and in fact $\|L\| \le 1$). Furthermore, $\inf_C \|\ \|$ is the (strictly positive) distance from x to D, and $\inf_C L = L(x) - \sup_D L$. Now let $z^* := L$. We will give another proof of this result using the "minimax technique" in Section 6. ▌

We now give three corollaries of Theorem 4.4 that we shall use later on. We leave the proofs of these as exercises. We shall use Corollary 4.5 (see also [46], Theorem 3.12, p. 64–65) and Corollary 4.6 in Lemma 32.1, and Corollary 4.7 (see also [46], Theorem 3.5, p. 59.) in Theorem 36.2.

Corollary 4.5. *If F is a nonempty closed convex subset of E then F is $w(E, E^*)$–closed.*

Corollary 4.6. *If F is a nonempty closed convex subset of E, C is a nonempty $w(E, E^*)$–compact convex subset of E and $F \cap C = \emptyset$ then $F - C$ is closed and* $\inf \|F - C\| > 0$.

Corollary 4.7. *If D is a subspace of E and*

$$z^* \in E^* \text{ and } \langle y, z^* \rangle = 0 \text{ for all } y \in D \quad \Longrightarrow \quad z^* = 0,$$

then D is dense in E.

We now give a minimax proof of a result that we shall need in Lemma 30.1 and Theorem 40.2. (Theorem 4.8 can also be established using the theory of locally convex spaces and the "bipolar theorem".)

Theorem 4.8. *Let C be a nonempty $w(E^*, E)$–compact convex subset of E^*, and define the sublinear functional S on E by $S(x) := \max\langle x, C\rangle$ $(x \in E)$. Let $x^* \in E^*$ and*

$$x^* \le S \text{ on } E.$$

Then $x^* \in C$.

Proof. Define $h\colon E \times C \mapsto \mathbb{R}$ by $h(x, y^*) := \langle x, y^* - x^* \rangle$. Then our assumption is exactly that

$$\inf_E \max_C h \ge 0.$$

The function h is linear on E, and $w(E^*, E)$–continuous and affine on C. Thus from the minimax theorem, Theorem 3.1,

$$\max_C \inf_E h \ge 0,$$

that is to say, there exists $y^* \in C$ such that

$$x \in E \quad \Longrightarrow \quad \langle x, y^* - x^* \rangle \ge 0.$$

This clearly implies that $x^* = y^*$, hence $x^* \in C$, as required. ▌

We now give the "extension form" of the Hahn-Banach theorem. The only place in these notes where we will use this result is in Theorem 16.10.

Theorem 4.9. *Let F be a subspace of E and $y^* \in F^*$. Then there exists $x^* \in E^*$ such that*

$$\|x^*\| = \|y^*\| \quad \text{and} \quad x^*|_F = y^*.$$

Proof. See Rudin, [46], Theorem 3.2, p. 56–57 for a conventional proof of this result. We will give another proof using the minimax technique in Section 6. ▌

Remark 4.10. One can easily deduce the following standard "separation theorem" (which we will not need) from Corollary 4.6 and Theorem 4.4, with $D := F - C$. *If F is a nonempty closed convex subset of E, C is a nonempty $w(E, E^*)$–compact convex subset of E and $F \cap C = \emptyset$ then there exists $z^* \in E^*$ such that $\sup_F z^* < \inf_C z^*$.* (See see Rudin, [46], Theorem 3.4, p. 58–59.)

We conclude this section by mentioning James's theorem, one of the most beautiful results in functional analysis: *if C is a nonempty bounded closed convex subset of E then C is $w(E, E^*)$–compact if, and only if, for all $x^* \in E^*$, there exists $x \in C$ such that $\langle x, x^* \rangle = \max_C x^*$.* James's theorem is not easy — we refer the reader to Pryce, [37] for a proof. We shall use the following special case in the construction of a couple of examples. This special case follows from Theorem 4.3.

Theorem 4.11. *E is reflexive if, and only if, for all $x^* \in E^*$ there exists $x \in E$ such that $\|x\| \le 1$ and $\langle x, x^* \rangle = \|x^*\|$.*

5. The minimax criterion for weak compactness in a Banach space

It is immediate from Theorem 3.1 (using the definition of $w(E, E^*)$ and the bilinearity of the function $\langle \cdot, \cdot \rangle$ on $E \times E^*$) that if B is a nonempty convex $w(E, E^*)$–compact subset of E and A is a nonempty convex subset of E^* then

$$\inf_{x^* \in A} \max_{x \in B} \langle x, x^* \rangle = \max_{x \in B} \inf_{x^* \in A} \langle x, x^* \rangle.$$

We shall prove in Theorem 5.1 that the converse of this result is true. Theorem 5.1 first appeared in our paper [48].

Theorem 5.1. *Let B be a nonempty bounded closed convex subset of E. Suppose that, for all nonempty bounded norm–closed convex subsets A of E^*,*

$$\sup_{x \in B} \inf_{x^* \in A} \langle x, x^* \rangle \ge \inf_{x^* \in A} \sup_{x \in B} \langle x, x^* \rangle. \tag{5.1.1}$$

Then B is $w(E, E^*)$–compact.

Proof. Let C be the $w(E^{**}, E^*)$ closure of $\widehat{B}$ in E^{**}, and x^{**} be an arbitrary element of C. The main step in the proof is to establish that,

$$\text{for all } \varepsilon > 0, \quad \text{there exists } v \in B \text{ such that} \quad \|x^{**} - \widehat{v}\| < \varepsilon. \tag{5.1.2}$$

Let us suppose that this has been done, and show how to deduce that B is $w(E, E^*)$–compact. As we have already oberved, $\widehat{B}$ is a norm–closed subset of E^{**}, thus it follows from (5.1.2) that $x^{**} \in \widehat{B}$. Since this holds for any

element x^{**} of C, we have shown that $C = \widehat{B}$, from which $\widehat{B}$ is $w(E^{**}, E^*)$–closed. Since $\widehat{B}$ is bounded, and consequently contained in some ball, we obtain from the fact that a closed subset of a compact set is compact and the Banach–Alaoglu theorem, Theorem 4.1, that $\widehat{B}$ is $w(E^{**}, E^*)$–compact. As we have also observed, $\widehat{\ }$ is a homeomorphism of E onto $\widehat{E}$ with respect to $w(E, E^*)$ and $w(E^{**}, E^*)$, consequently B is $w(E, E^*)$–compact, as required.
Now we give the details of the proof of (5.1.2). For all $x^* \in E^*$, the map $\langle x^*, \cdot\rangle$ is $w(E^{**}, E^*)$–continuous on E^{**}, consequently

$$x^* \in E^* \quad \Longrightarrow \quad \sup\langle B, x^*\rangle = \sup\langle x^*, \widehat{B}\rangle \geq \langle x^*, x^{**}\rangle. \tag{5.1.3}$$

Let Z be the unit ball of E^*. We can write $Z = A_1 \cup \ldots \cup A_m$ where, for all $i \in \{1, \ldots, m\}$, A_i is a nonempty bounded norm–closed convex subset of E^* and

$$\sup\langle A_i, x^{**}\rangle - \inf\langle A_i, x^{**}\rangle < \varepsilon. \tag{5.1.4}$$

Let $i \in \{1, \ldots, m\}$. From (5.1.1), (5.1.3) and (5.1.4),

$$\sup_{x\in B}\inf_{x^*\in A_i}\langle x, x^*\rangle \geq \inf_{x^*\in A_i}\sup\langle B, x^*\rangle \geq \inf\langle A_i, x^{**}\rangle > \sup\langle A_i, x^{**}\rangle - \varepsilon,$$

hence there exists $x_i \in B$ such that

$$\inf\langle x_i, A_i\rangle > \sup\langle A_i, x^{**}\rangle - \varepsilon,$$

from which

$$\sup\langle A_i, x^{**} - \widehat{x_i}\rangle \leq \sup\langle A_i, x^{**}\rangle - \inf\langle x_i, A_i\rangle < \varepsilon.$$

Since $Z = A_1 \cup \ldots \cup A_m$,

$$\sup_Z[(x^{**} - \widehat{x_1}) \wedge \cdots \wedge (x^{**} - \widehat{x_m})] < \varepsilon.$$

From Lemma 2.1(b) with $X = Z$ and $g_i := x^{**} - \widehat{x_i}$, there exist $\lambda_1, \ldots, \lambda_m \geq 0$ such that $\lambda_1 + \ldots + \lambda_m = 1$ and

$$\sup_Z[\lambda_1(x^{**} - \widehat{x_1}) + \cdots + \lambda_m(x^{**} - \widehat{x_m})] < \varepsilon.$$

Setting $v := \lambda_1 x_1 + \cdots + \lambda_m x_m \in B$,

$$\sup_Z(x^{**} - \widehat{v}) < \varepsilon,$$

i.e.,

$$\|x^{**} - \widehat{v}\| < \varepsilon.$$

This completes the proof of (5.1.2) and hence that of Theorem 5.1. ∎

6. Four examples of the "minimax technique" — Fenchel duality

In this section, we will give four examples of the "minimax technique". While all four problems can be solved using other techniques, we thought that it would be a good idea to isolate with simple examples the usefulness of the ability to "switch quantifiers". In later sections, we will have many examples where the minimax technique can be used to obtain new results.

The minimax technique assumes that we know the one–dimensional Hahn–Banach theorem, Corollary 1.2, the minimax theorem, Theorem 3.1, and the Banach–Alaoglu theorem, Theorem 4.1 or the extended Banach–Alaoglu theorem, Theorem 4.2. We first show how this technique can be used to prove two results that we have already given.

The first of these, admittedly contrived, is the extension form of the Hahn–Banach theorem as stated in Theorem 4.9. So we suppose that F is a subspace of E and $y^* \in F^*$, and we ask the question: *when does there exist* $x^* \in E^*$ *such that*

$$x^*|_F = y^* \quad \text{and} \quad \|x^*\| \le \|y^*\|. \tag{6.0.1}$$

Since F is a subspace of E, the statement "$x^*|_F = y^*$" is equivalent to

$$\text{for all } y \in F, \quad \langle y, x^* \rangle - \langle y, y^* \rangle \ge 0.$$

Let $B := \{x^* \in E^*\colon \|x^*\| \le \|y^*\|\}$ with the topology $w(E^*, E)$. From the Banach–Alaoglu theorem, Theorem 4.1, B is compact. Define $h\colon F \times B \mapsto \mathbb{R}$ by

$$h(y, x^*) := \langle y, x^* \rangle - \langle y, y^* \rangle.$$

So our question is equivalent to finding when

$$\max_B \inf_F h \ge 0.$$

The function h is linear on F, and affine and continuous on B. Thus, from the minimax theorem, Theorem 3.1, our question is equivalent to finding when

$$\inf_F \max_B h \ge 0,$$

that is to say,

$$\text{for all } y \in F, \quad \max_{x^* \in B} \langle y, x^* \rangle \ge \langle y, y^* \rangle.$$

Using the one–dimensional Hahn–Banach theorem, Corollary 1.2, this is equivalent to the statement

$$\text{for all } y \in F, \quad \|y\|\|y^*\| \ge \langle y, y^* \rangle.$$

Since this is always true, there always exists $x^* \in E^*$ satisfying (6.0.1).

We now apply the same technique to another result that we have already given, the separation theorem of Theorem 4.4. So we suppose that D is a nonempty convex subset of E and $x \in E$, and we ask the question: *when does there exist* $z^* \in E^*$ *such that*

$$\sup_D z^* < \langle x, z^* \rangle. \tag{6.0.2}$$

This is equivalent to the problem: *when do there exist* $M > 0$ *and* $\delta > 0$ *such that there exists* $z^* \in E^*$ *such that* $\|z^*\| \le M$ *and*

$$\text{for all } y \in D, \quad \langle x, z^* \rangle - \langle y, z^* \rangle \ge \delta.$$

We now use exactly the same argument as above with

$$B := \{z^* \in E^* \colon \|z^*\| \le M\}$$

and $h\colon D \times B \mapsto \mathbb{R}$ defined by

$$h(y, z^*) := \langle x, z^* \rangle - \langle y, z^* \rangle = \langle x - y, z^* \rangle.$$

It follows that our problem is equivalent to: *when do there exist* $M > 0$ *and* $\delta > 0$ *such that*

$$\text{for all } y \in D, \quad M\|x - y\| \ge \delta.$$

This happens exactly when $x \notin \overline{D}$. So there exists x^* satisfying (6.0.2) if, and only if, $x \notin \overline{D}$.

The above example shows how the minimax technique can be used to transform a problem on the existence of linear functionals into a problem on the existence of one or more real constants. We now give as exercises two further examples of this kind.

Example 6.1. Let A be a nonempty convex subset of E and $f\colon A \mapsto \mathbb{R}$ be convex. Prove that *there exists* $x^* \in E^*$ *such that* $x^* \le f$ *on* A

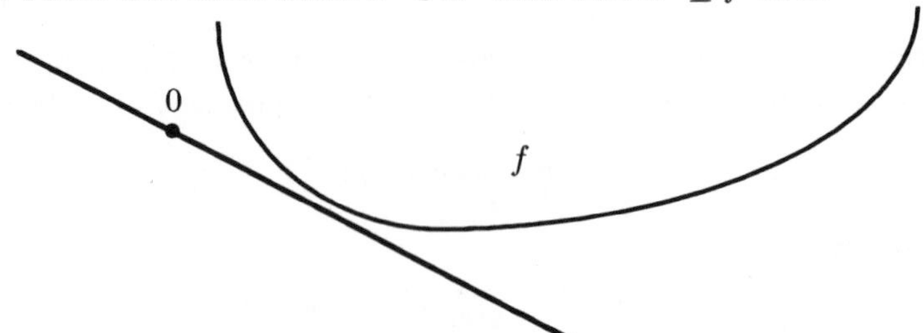

if, and only if, there exists $M \ge 0$ *such that*

$$\text{for all } x \in A, \quad f(x) + M\|x\| \ge 0.$$

Example 6.2. Let f_1: $E \mapsto \mathbb{R} \cup \{\infty\}$ and f_2: $E \mapsto \mathbb{R} \cup \{\infty\}$ be convex, $\operatorname{dom} f_1 \neq \emptyset$ and $\operatorname{dom} f_2 \neq \emptyset$. Prove that *there exist $z^* \in E^*$ and $\beta \in \mathbb{R}$ such that*

$$-f_1 \le z^* + \beta \le f_2 \quad \text{on} \quad E \tag{6.2.1}$$

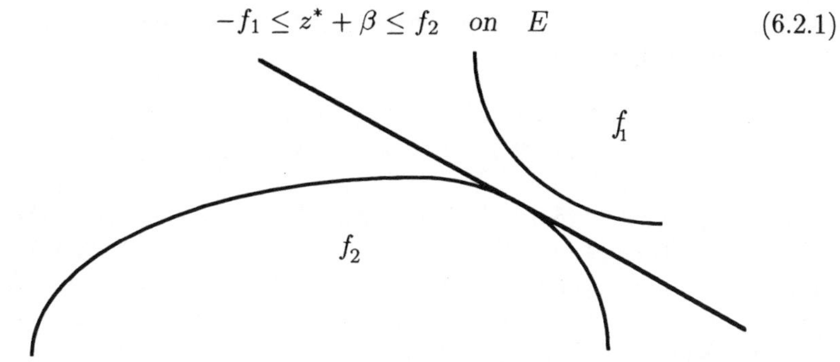

if, and only if,

$$\left.\begin{array}{l} \text{there exists } M \ge 0 \text{ such that,} \\ \quad \text{for all } x_1,\ x_2 \in E, \\ \qquad f_1(x_1) + f_2(x_2) + M\|x_1 - x_2\| \ge 0. \end{array}\right\} \tag{6.2.2}$$

We note that (6.2.1) can be split up into the two statements "$-z^* - f_1 \le \beta$ on E" and "$z^* - f_2 \le -\beta$ on E", that is to say "$f_1^*(-z^*) \le \beta$" and "$f_2^*(z^*) \le -\beta$", where the *Fenchel conjugate* f^* is defined by

$$f^*(x^*) := \sup_E (x^* - f).$$

It follows that (6.2.2) is equivalent to:

$$\text{there exists } z^* \in E^* \text{ such that} \quad f_1^*(-z^*) + f_2^*(z^*) \le 0. \tag{6.2.3}$$

This is an old condition in convex analysis, due to Fenchel in the finite dimensional case. Rockafeller proved in [40] that *if $-f_1 \le f_2$ on E and there is a point in $\operatorname{dom} f_1 \cap \operatorname{dom} f_2$ at which one of f_1 and f_2 is continuous then (6.2.3) holds.* In fact, one can show that (6.2.2) is true under these conditions (we leave this an an exercise — see Lemma 28.1 for a hint how to proceed), thus providing a minimax proof of Rockafellar's result. Attouch–Brézis have proved (this follows from [1], Corollary 2.3, p. 131–132) that

$$\left.\begin{array}{l} \text{if } f_1 \text{ and } f_2 \text{ are lower semicontinuous, } -f_1 \le f_2 \text{ on } E \text{ and} \\ \bigcup_{\lambda>0} \lambda(\operatorname{dom} f_1 - \operatorname{dom} f_2) \quad \text{is a closed subspace of } E \text{ then} \\ \text{there exists } z^* \in E^* \text{ such that} \quad f_1^*(-z^*) + f_2^*(z^*) \le 0. \end{array}\right\} \tag{6.2.4}$$

In fact, one can show using Baire's theorem that (6.2.2) is also true under these conditions (see Lemma 14.1), thus providing a minimax proof of (6.2.4) also. However, we emphasize that (6.2.2) is a *necessary* as well as a sufficient condition for (6.2.3) and also that (6.2.2) does not require any continuity or lower semicontinuity conditions on f_1 or f_2.

Remark 6.3. We indicate how the result of Example 6.1 and an inequality that we shall establish later lead to two more classical results from convex analysis. The first is the following: *Let* $g\colon E \mapsto \mathbb{R} \cup \{\infty\}$ *be convex and lower semicontinuous and* $\operatorname{dom} g \neq \emptyset$. *Then* $\operatorname{dom} g^* \neq \emptyset$. To see this, let $\lambda < g(0)$. We first show that

$$\left.\begin{aligned} &\text{there exists } M \geq 0 \text{ such that,}\\ &\quad\text{for all } x \in E, \quad g(x) - \lambda + M\|x\| \geq 0. \end{aligned}\right\} \tag{6.3.1}$$

If $\inf_E g \geq \lambda$ then (6.3.1) follows with $M := 0$. If, on the other hand, $\inf_E g < \lambda$ then, from inequality (29.4.6) with $S := \|\ \|$, there exists $K \in (0, \infty)$ such that $\inf_E(g + K\|\ \|) \geq \lambda$, and (6.3.1) follows with $M := K$. We now derive from Example 6.1 that there exists $x^* \in E^*$ such that $x^* \leq g - \lambda$ on E, which implies that

$$g^*(x^*) \leq -\lambda, \tag{6.3.2}$$

hence $\operatorname{dom} g^* \neq \emptyset$, as required.

Now suppose that $f\colon E \mapsto \mathbb{R} \cup \{\infty\}$ is convex and lower semicontinuous, and $\operatorname{dom} f \neq \emptyset$. Clearly, $f^*\colon E \mapsto \mathbb{R} \cup \{\infty\}$ is convex and lower semicontinuous and, from what we have proved, $\operatorname{dom} f^* \neq \emptyset$. We define ${}^*f^*\colon E \mapsto \mathbb{R}\cup\{\infty\}$ (the *restricted biconjugate* of f) by

$${}^*f^*(x) := \sup_{E^*}(\widehat{x} - f^*).$$

Now ${}^*f^* \leq f$ on E (exercise!). We now establish the formula (associated by various authors with the names of Legendre, Fenchel, Moreau and Hormander) that

$${}^*f^* = f \text{ on } E.$$

To see this, let $y \in E$ and $\lambda < f(y)$. Define $g\colon E \mapsto \mathbb{R} \cup \{\infty\}$ by $g(x) := f(x + y) \quad (x \in E)$. Then (exercise!)

$$\text{for all } x^* \in E^*,\ g^*(x^*) = f^*(x^*) - \langle y, x^* \rangle.$$

By hypothesis, $\lambda < g(0)$ so, from (6.3.2), there exists $x^* \in E^*$ such that

$$f^*(x^*) - \langle y, x^* \rangle = g^*(x^*) \leq -\lambda.$$

Thus

$${}^*f^*(y) \geq \langle y, x^* \rangle - f^*(x^*) \geq \lambda.$$

Letting $\lambda \to f(y)$, we obtain that ${}^*f^*(y) \geq f(y)$, as required.

7. The perfect square trick and the fg–theorem

The main result of this section is the fg–theorem, Theorem 7.2, which we shall use in Lemmas 10.1, 20.1 and 27.4. This result is interesting in that is uses the minimax theorem twice, first in a scalar form to produce the bound β, and then in a vector form in which this bound is used to establish compactness. Before giving the specifics of the fg–theorem, we give a simple result that we shall call the "perfect square trick".

Lemma 7.1. *If $x \in E$ and $x^* \in E^*$ then $\|x\|^2 + \|x^*\|^2 + 2\langle x, x^*\rangle \ge 0$.*

Proof. $\|x\|^2+\|x^*\|^2+2\langle x,x^*\rangle \ge \|x\|^2+\|x^*\|^2-2\|x\|\|x^*\| = (\|x\|-\|x^*\|)^2$. ∎

Theorem 7.2. *Let A be a nonempty convex subset of a vector space, F be a Banach space, $f\colon A \mapsto \mathbb{R}$ be convex and $g\colon A \mapsto F$ be affine. Then (7.2.1)$\Longleftrightarrow$(7.2.2).*

$$a \in A \quad \Longrightarrow \quad f(a) + \|g(a)\|^2 \ge 0. \tag{7.2.1}$$

$$\left.\begin{array}{l}\text{There exists } y^* \in F^* \text{ such that}\\ \qquad a \in A \quad \Longrightarrow \quad f(a) - 2\langle g(a), y^*\rangle - \|y^*\|^2 \ge 0.\end{array}\right\} \tag{7.2.2}$$

Proof. ($\Longrightarrow$) Let $a_0 \in A$, and n be an integer such that $n \ge \|g(a_0)\|$. Put

$$A_n := \{a \in A\colon\ \|g(a)\| \le n\}.$$

Since $a_0 \in A_n$, A_n is not empty, and A_n is clearly convex. Define $h\colon A \times [0,\infty) \mapsto \mathbb{R}$ by

$$h(a,\beta) := f(a) + 2\beta\|g(a)\| - \beta^2.$$

Using (7.2.1), we have

$$\inf_{A_n}\max_{[0,n]} h \ge \inf_{a \in A_n} h(a, \|g(a)\|) \ge 0.$$

The function h is convex on A_n, and concave and continuous on $[0,n]$. Since $[0,n]$ is compact, from the minimax theorem, Theorem 3.1,

$$\max_{[0,n]}\inf_{A_n} h \ge 0,$$

from which

$$C_n := \bigcap_{a \in A_n} \{\beta \in [0,\infty)\colon\ h(a,\beta) \ge 0\} \ne \emptyset.$$

C_n is clearly compact. Further, the sets C_n decrease as n increases. Consequently,

$$\bigcap_{n \ge \|g(a_0)\|} C_n \ne \emptyset.$$

Since

$$A = \bigcup_{n \geq \|g(a_0)\|} A_n,$$

it now follows that

$$\left.\begin{array}{c}\text{there exists } \beta \geq 0 \text{ such that} \\ a \in A \quad \Longrightarrow \quad f(a) + 2\beta\|g(a)\| - \beta^2 \geq 0.\end{array}\right\} \tag{7.2.3}$$

Now define $h\colon\ A \times F^* \mapsto \mathbb{R}$ by

$$h(a, y^*) := f(a) - 2\langle g(a), y^* \rangle - \|y^*\|^2.$$

Let $B := \{y^* \in F^*\colon\ \|y^*\| \leq \beta\}$, with the topology $w(F^*, F)$. From the Banach–Alaoglu theorem, Theorem 4.1, B is compact. Let $a \in A$. From the one–dimensional Hahn–Banach theorem, Corollary 1.2, we can find $y^* \in B$ such that $\langle g(a), y^* \rangle = -\beta\|g(a)\|$. From (7.2.3),

$$\begin{aligned} h(a, y^*) &= f(a) - 2\langle g(a), y^* \rangle - \|y^*\|^2 \\ &\geq f(a) + 2\beta\|g(a)\| - \beta^2 \\ &\geq 0. \end{aligned}$$

Thus

$$\inf_A \max_B h \geq 0.$$

Since A and B are convex, and h is convex on A and concave and upper semicontinuous on B (exercise!), from the minimax theorem, Theorem 3.1,

$$\max_B \inf_A h \geq 0,$$

which gives (7.2.2).

($\Longleftarrow$) This is immediate from the perfect square trick, Lemma 7.1, with $E := F$, $x := g(a)$ and $x^* := y^*$. ∎

Remark 7.3. One can prove the following more abstract version of the implication ((7.2.1)$\Longrightarrow$(7.2.3)) (exercise!): *Let A be a nonempty convex subset of a vector space and $h\colon\ A \times [0, \infty) \mapsto \mathbb{R}$ be convex on A, and concave and upper semicontinuous on $[0, \infty)$. Suppose that $k\colon\ A \mapsto [0, \infty)$ is convex,*

$$a \in A \quad \Longrightarrow \quad h(a, k(a)) \geq 0$$

and

there exists $a_0 \in A$ such that $\{\beta \in [0, \infty)\colon\ h(a_0, \beta) \geq 0\}$ is bounded.

Then there exists $\beta \geq 0$ such that

$$a \in A \quad \Longrightarrow \quad h(a, \beta) \geq 0.$$

II. Multifunctions

8. Multifunctions, monotonicity and maximality

We now introduce some general notation for "multifunctions" or "set–valued maps". If Y is a nonempty set, we write 2^Y for the power set of Y, the set of all subsets of Y. If $S\colon X \mapsto 2^Y$, we write

$$G(S) := \{(x, y)\colon\ x \in X,\ y \in Sx\}.$$

$G(S)$ is the "graph" of S. We shall always suppose that $G(S) \neq \emptyset$ — we shall emphasize this by saying that S is *nontrivial*. We write

$$D(S) := \{x \in X\colon\ Sx \neq \emptyset\}.$$

$D(S)$ is the "domain" of S. We write

$$R(S) := \{y\colon \quad \text{there exists } x \in X \quad \text{such that} \quad y \in Sx\} = \textstyle\bigcup_{x \in X} Sx.$$

$R(S)$ is the "range" of S. Finally, if $S\colon X \mapsto 2^Y$, we define $S^{-1}\colon Y \mapsto 2^X$ by

$$S^{-1}y := \{x \in X\colon\ Sx \ni y\}.$$

S^{-1} is the "inverse" of S. Obviously $D(S^{-1}) = R(S)$ and $R(S^{-1}) = D(S)$. We point to the books [3] by Aubin, [4] by Aubin–Frankowska and [19] by Deimling as general references on multifunctions. We shall be concerned here with multifunctions from one Banach space into another, in which case additional operations can be defined. If $S_1\colon E \mapsto 2^F$ and $S_2\colon E \mapsto 2^F$ are nontrivial, we define $S_1 + S_2\colon E \mapsto 2^F$ (with $D(S_1 + S_2) = D(S_1) \cap D(S_2)$) by

$$(S_1 + S_2)x := S_1x + S_2x \quad (x \in E), \tag{8.0.1}$$

where $S_1x + S_2x$ is the "Minkowski sum" $\{y_1 + y_2\colon\ y_1 \in S_1x,\ y_2 \in S_2x\}$. As a special case of this, if $S\colon E \mapsto 2^F$ and $y \in F$ we define $S + y\colon E \mapsto 2^F$ with $D(S + y) = D(S)$ by

$$(S + y)x := Sx + y \quad (x \in E).$$

Obviously, $R(S+y) = R(S)+y$. The operation described above is sometimes referred to as "translation in F". There is also an operation of "translation in E", which we now describe. Let $S\colon E \mapsto 2^F$ and $z \in E$. Write $T := (S^{-1} - z)^{-1}$. Then,

$$\text{for all } x \in E, \quad Tx = S(x+z).$$

Here, $D(T) = D(S) - z$ and $R(T) = R(S)$.

Let $S\colon E \mapsto 2^{E^*}$. S is said to be *monotone* if

$$(x, x^*) \text{ and } (y, y^*) \in G(S) \quad \Longrightarrow \quad \langle x - y, x^* - y^* \rangle \geq 0.$$

We point to the notes [35] and the book [34] by Phelps, and the book [62] by Zeidler as general references on monotone multifunctions. S is said to be *maximal monotone* if S is monotone, and S has no proper monotone extension. This is equivalent to the statement (exercise!):

$$(z, z^*) \in E \times E^* \text{ and } \inf_{(s,s^*)\in G(S)} \langle s - z, s^* - z^* \rangle \geq 0 \quad \Longrightarrow \quad (z, z^*) \in G(S).$$

It will be convenient to have the more quantitative version of this that is provided by Lemma 8.1. (a) follows by taking $(s, s^*) := (z, z^*)$, (b) is immediate from the definition of maximal monotonicity, and (c) follows from (a) and (b).

Lemma 8.1. *Let $S\colon E \mapsto 2^{E^*}$ be maximal monotone.*
(a) If $(z, z^*) \in G(S)$ then $\inf_{(s,s^*)\in G(S)} \langle s - z, s^* - z^* \rangle = 0$.
(b) If $(z, z^*) \in E \times E^* \setminus G(S)$ then $\inf_{(s,s^*)\in G(S)} \langle s - z, s^* - z^* \rangle < 0$.
(c) For all $(z, z^*) \in E \times E^*$, $\inf_{(s,s^*)\in G(S)} \langle s - z, s^* - z^* \rangle \leq 0$.

In order to simplify matters a little, if $(z, z^*) \in E \times E^*$ and $G \subset E \times E^*$, we shall say that (z, z^*) is *monotonically related* to G when

$$\inf_{(s,s^*)\in G} \langle s - z, s^* - z^* \rangle \geq 0.$$

We now give some examples of maximal monotone multifunctions. The first one we consider is that of *positive linear operators*. Let $T\colon E \mapsto E^*$ be linear and

$$x \in E \quad \Longrightarrow \quad \langle x, Tx \rangle \geq 0.$$

Then T *is a (single valued) maximal monotone operator*. More precisely, *the multifunction S defined by $Sx := \{Tx\}$ is maximal monotone.* The monotonicity is easy to see. To prove the maximality, suppose that $(z, z^*) \in E \times E^*$ is monotonically related to $G(S)$. Then

$$\inf_{y\in E} \langle y - z, Ty - z^* \rangle \geq 0.$$

Let $x \in E$, $\lambda \in \mathbb{R}$, and put $y := z + \lambda x$ (exercise!). As a special case of the above, we mention *skew linear operators*. These are linear operators $T\colon E \mapsto E^*$ such that

$$x \in E \quad \Longrightarrow \quad \langle x, Tx \rangle = 0.$$

See the papers [5] and [6] by Bauschke–Borwein and the paper [36] by Phelps–Simons for recent work on positive linear operators.

The second example that we consider is that of *subdifferentials*. We shall write $\mathcal{PCLSC}(E)$ for the set of all convex lower semicontinuous functions $f\colon E \mapsto \mathbb{R} \cup \{\infty\}$ such that $\operatorname{dom} f \neq \emptyset$. (The "$\mathcal{P}$" stands for "proper", which is the adjective frequently used to denote the fact that a function is somewhere finite.) If $f \in \mathcal{PCLSC}(E)$ and $x \in E$, the *subdifferential* of f at x is defined by

$$\partial f(x) := \{z^* \in E^*\colon \quad y \in E \Longrightarrow f(x) + \langle y - x, z^*\rangle \leq f(y)\}.$$

Then $\partial f\colon E \mapsto 2^{E^*}$ *is maximal monotone*. The monotonicity is easy to see. The maximality is not easy, and is *Rockafeller's maximal monotonicity theorem*. (See Chapter VII for a proof of this, and other properties of subdifferentials.) It is easy to see in this situation that $D(\partial f) \subset \operatorname{dom} f$ (exercise!), however this inclusion may be proper: let $E := \mathbb{R}$ and $f\colon \mathbb{R} \mapsto \mathbb{R} \cup \{\infty\}$ be defined by

$$f(x) := \begin{cases} -\sqrt{1-x^2}, & \text{if } x \in [-1,1]; \\ \infty, & \text{otherwise;} \end{cases}$$

then $D(\partial f) = (-1,1)$ but $\operatorname{dom} f = [-1,1]$. The Brøndsted–Rockafellar theorem (see Corollary 29.2) establishes a close connection between $D(\partial f)$ and $\operatorname{dom} f$: *Let $f \in \mathcal{PCLSC}(E)$, $\alpha, \beta > 0$, $y \in \operatorname{dom} f$ and $f(y) \leq \inf_E f + \alpha\beta$. Then there exist $x \in E$ and $x^* \in \partial f(x)$ such that $\|x - y\| \leq \alpha$, $f(x) \leq f(y)$ and $\|x^*\| \leq \beta$.* In particular, *$D(\partial f)$ is dense in $\operatorname{dom} f$.* Incidentally, if T is linear, skew and nonzero then T is not a subdifferential so, provided that E has dimension > 1, there always exist maximal monotone multifunctions that are not subdifferentials. (If $E = \mathbb{R}$ then every maximal monotone multifunction on E is a subdifferential (exercise!).)

The final example that we mention here is that of the *normality multifunction*. Let C be a nonempty closed convex subset of E and $N_C\colon E \mapsto 2^{E^*}$ be defined by

$$(x, x^*) \in G(N_C) \iff x \in C \text{ and } \langle x, x^*\rangle = \max_C x^*. \tag{8.1.1}$$

Then N_C *is maximal monotone*. Again, the monotonicity is easy to see. The maximality can be seen in two ways. First, if we define $I_C\colon E \mapsto \mathbb{R} \cup \{\infty\}$ to be the "indicator function of C", that is to say

$$I_C(x) := \begin{cases} 0, & \text{if } x \in C; \\ \infty, & \text{otherwise;} \end{cases}$$

then $N_C = \partial I_C$. Since I_C is convex and lower semicontinuous, it follows from the result of Rockafellar mentioned above that N_C is maximal monotone. Alternatively, one can proceed directly from the definition of N_C and use the consequence of the Bishop–Phelps theorem in Phelps, [34], Proposition 3.20, p. 49 that *C is the intersection of the closed half-spaces defined by its supporting hyperplanes.*

We point out finally that the Bishop–Phelps theorem and the Brøndsted–Rockafellar theorem mentioned above were both precursors (and are consequences of) Ekeland's variational principle (see Theorem 29.1.)

9. The "big convexification"

Much of this section first appeared (with a different notation) in the paper [18] by Coodey–Simons. We write $\mathbb{R}^{(E\times E^*)}$ for the direct sum of $E \times E^*$ copies of $\mathbb{R}$, that is the set of functions μ: $E \times E^* \mapsto \mathbb{R}$ such that

$$\{(s, s^*) \in E \times E^*\colon\ \mu(s, s^*) \neq 0\} \quad \text{is finite.}$$

$\mathbb{R}^{(E\times E^*)}$ is a vector space. If $(y, y^*) \in E \times E^*$ then $\delta_{(y,y^*)} \in \mathbb{R}^{(E\times E^*)}$, where $\delta_{(y,y^*)}$ is defined by

$$\delta_{(y,y^*)}(s, s^*) := \begin{cases} 1, & \text{if } (s, s^*) = (y, y^*); \\ 0, & \text{otherwise.} \end{cases}$$

We can also think of $\mathbb{R}^{(E\times E^*)}$ as the set of signed measures on $E \times E^*$ with finite support. Then $\delta_{(y,y^*)}$ corresponds to the point mass at (y, y^*). Any nonempty subset of $E \times E^*$ has a "big convexification" in the following sense: let $\emptyset \neq G \subset E \times E^*$, and write $\mathcal{CO}(G)$ for the convex hull in $\mathbb{R}^{(E\times E^*)}$ of $\{\delta_{(y,y^*)}\colon\ (y, y^*) \in G\}$. Explicitly, if $\mu \in \mathbb{R}^{(E\times E^*)}$ then $\mu \in \mathcal{CO}(G)$ if, and only if

$$\mu \geq 0 \text{ on } E \times E^*, \quad \mu(s, s^*) > 0 \Longrightarrow (s, s^*) \in G \quad \text{and} \quad \sum_{(s,s^*)\in G} \mu(s, s^*) = 1.$$

Continuing the "measure theory" analogy introduced above, we can think of $\mathcal{CO}(G)$ as the set of probability measures on $E \times E^*$ with finite support contained in G.

We now introduce the three linear operators p: $\mathbb{R}^{(E\times E^*)} \mapsto E$, q: $\mathbb{R}^{(E\times E^*)} \mapsto E^*$ and r: $\mathbb{R}^{(E\times E^*)} \mapsto \mathbb{R}$, defined by

$$p(\mu) := \textstyle\sum_{(s,s^*)\in E\times E^*} \mu(s, s^*)s,$$
$$q(\mu) := \textstyle\sum_{(s,s^*)\in E\times E^*} \mu(s, s^*)s^*$$

and

$$r(\mu) := \textstyle\sum_{(s,s^*)\in E\times E^*} \mu(s, s^*)\langle s, s^*\rangle.$$

We shall have frequent occasion to use the fact that,

$$\left.\begin{array}{l} \text{for all } (y, y^*) \in E \times E^*, \\ \qquad p(\delta_{(y,y^*)}) = y, \quad q(\delta_{(y,y^*)}) = y^* \quad \text{and} \quad r(\delta_{(y,y^*)}) = \langle y, y^*\rangle. \end{array}\right\}$$

A subset M of $E \times E^*$ is *monotone* if

$$(x, x^*) \text{ and } (y, y^*) \in M \quad \Longrightarrow \quad \langle x - y, x^* - y^* \rangle \geq 0.$$

The *pqr*–lemma, Lemma 9.1,is going to be basic in everything that follows. In it, we give a characterization of the monotone subsets of $E \times E^*$ in terms of p, q and r. The manipulations contained in Lemma 9.1 are part of the folklore of monotonicity.

Lemma 9.1. *Let M be a nonempty subset of $E \times E^*$. Then M is monotone if, and only if,*

$$\mu \in \mathcal{CO}(M) \quad \Longrightarrow \quad r(\mu) \geq \langle p(\mu), q(\mu) \rangle.$$

Proof. Suppose first that M is a monotone subset of $E \times E^*$. Let $(s_1, s_1^*), \ldots,$ (s_m, s_m^*) be an enumeration of those elements (s, s^*) of M for which $\mu(s, s^*) >$ 0, and write α_i for $\mu(s_i, s_i^*)$. Then, with the summations going from 1 to m,

$$\begin{aligned} r(\mu) - \langle p(\mu), q(\mu) \rangle &= \textstyle\sum_i \alpha_i \langle s_i, s_i^* \rangle - \langle \sum_i \alpha_i s_i, \sum_i \alpha_i s_i^* \rangle \\ &= \textstyle\sum_{i,j} \alpha_i \alpha_j \langle s_i, s_i^* \rangle - \sum_{i,j} \alpha_i \alpha_j \langle s_i, s_j^* \rangle \\ &= \textstyle\sum_{i,j} \alpha_i \alpha_j \langle s_i, s_i^* - s_j^* \rangle \\ &= \textstyle\sum_{i<j} \alpha_i \alpha_j \langle s_i, s_i^* - s_j^* \rangle + \sum_{j<i} \alpha_i \alpha_j \langle s_i, s_i^* - s_j^* \rangle \\ &= \textstyle\sum_{i<j} \alpha_i \alpha_j \langle s_i, s_i^* - s_j^* \rangle + \sum_{i<j} \alpha_i \alpha_j \langle s_j, s_j^* - s_i^* \rangle \\ &= \textstyle\sum_{i<j} \alpha_i \alpha_j \langle s_i - s_j, s_i^* - s_j^* \rangle \geq 0. \end{aligned}$$

We leave the proof of the converse as an exercise. ▌

Corollary 9.2 contains two consequences of the *pqr*–lemma that will be useful in applications — Corollary 9.2(a) in Lemmas 10.2 and 27.5, and Corollary 9.2(b) in Lemmas 20.3 and 21.1.

Corollary 9.2. *(a) If M is a nonempty monotone subset of $E \times E^*$ then*

$$\mu \in \mathcal{CO}(M) \quad \Longrightarrow \quad 2r(\mu) + \|p(\mu)\|^2 + \|q(\mu)\|^2 \geq 0. \tag{9.2.1}$$

(b) If M_1 and M_2 are nonempty monotone subsets of $E \times E^*$ then

$$\left.\begin{aligned} &(\mu_1, \mu_2) \in \mathcal{CO}(M_1) \times \mathcal{CO}(M_2) \quad \Longrightarrow \\ &\qquad 2r(\mu_1 + \mu_2) + 2\|p(\mu_2 - \mu_1)\| \|q(\mu_2)\| \\ &\qquad\qquad + \|p(\mu_1)\|^2 + \|q(\mu_1 + \mu_2)\|^2 \geq 0. \end{aligned}\right\} \tag{9.2.2}$$

Proof. (a) Using the *pqr*–lemma, Lemma 9.1, and the perfect square trick, Lemma 7.1, with $x := p(\mu)$ and $x^* := q(\mu)$,

$$\begin{aligned} 2r(\mu) + \|p(\mu)\|^2 + \|q(\mu)\|^2 &\geq \|p(\mu)\|^2 + \|q(\mu)\|^2 + 2\langle p(\mu), q(\mu) \rangle \\ &\geq 0. \end{aligned}$$

(b) Using arguments similar to those above,

$$\begin{aligned}2r(\mu_1+\mu_2)+\|p(\mu_1)\|^2+\|q(\mu_1+\mu_2)\|^2 &\geq 2r(\mu_1)+2r(\mu_2)-2\langle p(\mu_1),q(\mu_1+\mu_2)\rangle\\ &\geq 2\langle p(\mu_1),q(\mu_1)\rangle+2\langle p(\mu_2),q(\mu_2)\rangle-2\langle p(\mu_1),q(\mu_1+\mu_2)\rangle\\ &= 2\langle p(\mu_2-\mu_1),q(\mu_2)\rangle.\end{aligned}$$

Consequently,

$$\begin{aligned}2r(\mu_1+\mu_2)+2\|p(\mu_2-\mu_1)\|\|q(\mu_2)\|+\|p(\mu_1)\|^2+\|q(\mu_1+\mu_2)\|^2 &\geq 2\|p(\mu_2-\mu_1)\|\|q(\mu_2)\|+2\langle p(\mu_2-\mu_1),q(\mu_2)\rangle\\ &\geq 0. \blacksquare\end{aligned}$$

10. Criteria for maximal monotonicity in reflexive spaces

In this section we assume that E is reflexive. The main result is the "perfect square" criterion for maximality, Theorem 10.3, in which we prove that *if M is a monotone subset of $E \times E^*$, then M is maximal monotone if, and only if, for all $(w,w^*) \in E \times E^*$, there exists $(x,x^*) \in M$ such that*

$$\|x-w\|^2+\|x^*-w^*\|^2+2\langle x-w,x^*-w^*\rangle=0.$$

We deduce from this in Theorem 10.7 one direction of Rockafellar's "surjectivity theorem", that *if $S\colon E \mapsto 2^{E^*}$ is maximal monotone then*

$$R(S+J)=E^*.$$

We point out Theorem 10.7 does not assume that E has been renormed to have any special properties. We shall give generalizations of some of the results in this section to the nonreflexive case in Section 27.

We start by using the minimax technique in the form of the fg–theorem, Theorem 7.2, to prove an equivalence for *arbitrary* nonempty subsets of $E \times E^*$. We note that (10.1.1) is identical with (9.2.1), except that M has been replaced by G. The analysis in this section is derived from results that first appeared in our paper [54], however the fg–theorem enables us to give much simpler proofs.

Lemma 10.1. *Let E be reflexive and $\emptyset \neq G \subset E \times E^*$. Then the conditions (10.1.1) and (10.1.2) are equivalent:*

$$\mu \in \mathcal{CO}(G) \quad\Longrightarrow\quad 2r(\mu)+\|p(\mu)\|^2+\|q(\mu)\|^2 \geq 0. \tag{10.1.1}$$

$$\left.\begin{aligned}&\text{There exists } (x,x^*) \in E \times E^* \text{ such that}\\ &\quad (s,s^*) \in G \quad\Longrightarrow\\ &\qquad\qquad 2\langle s-x,s^*-x^*\rangle \geq \|x^*\|^2+\|x\|^2+2\langle x,x^*\rangle.\end{aligned}\right\} \tag{10.1.2}$$

Proof. We shall establish the equivalence of (10.1.1) and (10.1.2) by proving their equivalence with the intermediate conditions (10.1.3) and (10.1.4) below:

$$\left.\begin{array}{l}\text{There exists } (x,x^*)\in E\times E^* \text{ such that}\\ \quad \mu\in\mathcal{CO}(G) \quad\Longrightarrow\\ \qquad 2r(\mu)-2\langle p(\mu),x^*\rangle-2\langle x,q(\mu)\rangle-\|x\|^2-\|x^*\|^2\ge 0.\end{array}\right\} \tag{10.1.3}$$

$$\left.\begin{array}{l}\text{There exists } (x,x^*)\in E\times E^* \text{ such that}\\ \quad (s,s^*)\in G \quad\Longrightarrow\\ \qquad 2\langle s,s^*\rangle-2\langle s,x^*\rangle-2\langle x,s^*\rangle-\|x\|^2-\|x^*\|^2\ge 0.\end{array}\right\} \tag{10.1.4}$$

((10.1.1)$\Longleftrightarrow$(10.1.3)) We write $A:=\mathcal{CO}(G)$, $F:=E\times E^*$ with

$$\|(x,x^*)\|:=\sqrt{\|x\|^2+\|x^*\|^2}$$

and, for all $\mu\in A$,

$$f(\mu):=2r(\mu)\quad\text{and}\quad g(\mu):=\big(p(\mu),q(\mu)\big).$$

Then (10.1.1) reduces to (7.2.1). It follows from the fg–theorem, Theorem 7.2, that (10.1.1) is equivalent to:

$$\left.\begin{array}{l}\text{there exists } y^*\in F^* \text{ such that}\\ \qquad \mu\in\mathcal{CO}(G)\quad\Longrightarrow\quad 2r(\mu)-2\langle g(\mu),y^*\rangle-\|y^*\|^2\ge 0,\end{array}\right\}$$

which is equivalent to (10.1.3) since any element y^* of F^* can be written in the form $(x^*,\widehat{x})$ for some $(x,x^*)\in E\times E^*$, and $\|y^*\|=\sqrt{\|x\|^2+\|x^*\|^2}$.

((10.1.3)$\Longleftrightarrow$(10.1.4)) If (10.1.3) is satisfied then (10.1.4) follows by restricting μ to the values $\delta_{(s,s^*)}$. If, conversely, (10.1.4) is satisfied and $\mu\in\mathcal{CO}(G)$ then (10.1.3) follows by multiplying the left hand side of the inequality in (10.1.4) by $\mu(s,s^*)$ and summing up over all $(s,s^*)\in G$.

((10.1.4)$\Longleftrightarrow$(10.1.2)) This can be seen by rearranging the terms and adding $\pm 2\langle x,x^*\rangle$ to each side. This completes the proof of Lemma 10.1. ∎

Note that we introduce the term $2\langle x,x^*\rangle$ *after* the minimax theorem has been used in the proof that (10.1.1) implies (10.1.2) above, since this term is not generally semicontinuous on $E\times E^*$ with respect to the product of the topologies $w(E,E^*)$ and $w(E^*,E)$.

Lemma 10.2. *Let E be reflexive and M be a nonempty monotone subset of $E\times E^*$. Then:*
(a) There exists $(x,x^*)\in E\times E^*$ such that

$$(s,s^*)\in M\quad\Longrightarrow\quad 2\langle s-x,s^*-x^*\rangle\ge\|x^*\|^2+\|x\|^2+2\langle x,x^*\rangle. \tag{10.2.1}$$

(b) Suppose now that M is a maximal monotone *subset of $E \times E^*$. Then there exists $(x, x^*) \in M$ such that*

$$\|x\|^2 + \|x^*\|^2 + 2\langle x, x^*\rangle = 0.$$

Proof. We note that (10.2.1) is identical with (10.1.2), except that G has been replaced by M. Thus (a) immediate from Corollary 9.2(a) and Lemma 10.1. Now let $(x, x^*) \in E \times E^*$ be as in (a). From the perfect square trick, Lemma 7.1,

$$(s, s^*) \in M \implies 2\langle s - x, s^* - x^*\rangle \geq 0.$$

Since M is *maximal* monotone, $(x, x^*) \in M$, and (b) then follows from the perfect square trick again by substituting $(s, s^*) := (x, x^*)$ in (10.2.1). ▌

We now come to the perfect square criterion for maximality.

Theorem 10.3. *Let E be reflexive and M be a monotone subset of $E \times E^*$. Then M is maximal monotone $\iff$ for all $(w, w^*) \in E \times E^*$, there exists $(x, x^*) \in M$ such that*

$$\|x - w\|^2 + \|x^* - w^*\|^2 + 2\langle x - w, x^* - w^*\rangle = 0. \tag{10.3.1}$$

Proof. ($\Longrightarrow$) We apply Lemma 10.2(b), with M replaced by $M - (w, w^*) \subset E \times E^*$, which is also maximal monotone.
($\Longleftarrow$) Let $(w, w^*) \in E \times E^*$ and

$$(x, x^*) \in M \implies \langle x - w, x^* - w^*\rangle \geq 0.$$

Choose $(x, x^*) \in M$ as in (10.3.1). Then

$$\|x - w\|^2 + \|x^* - w^*\|^2 \leq 0,$$

and so $(w, w^*) = (x, x^*) \in M$. Thus M is maximal monotone. ▌

We next deduce from Theorem 10.3 the "negative alignment" criterion for maximality.

Corollary 10.4. *Let E be reflexive and M be a monotone subset of $E \times E^*$. Then M is maximal monotone $\iff$ for all $(w, w^*) \in E \times E^* \setminus M$, there exists $(x, x^*) \in M$ such that*

$$x \neq w, \; x^* \neq w^* \quad \text{and} \quad \langle x - w, x^* - w^*\rangle = -\|x - w\|\|x^* - w^*\|. \tag{10.4.1}$$

Proof. ($\Longrightarrow$) Let $(w, w^*) \in E \times E^* \setminus M$. Choose $(x, x^*) \in M$ as in (10.3.1). Clearly, either $x \neq w$ or $x^* \neq w^*$ (or both!). It follows from the proof of the perfect square trick, Lemma 7.1, that $\|x - w\| = \|x^* - w^*\|$. So, in fact $x \neq w$ *and* $x^* \neq w^*$. Using (10.3.1) and the proof of the perfect square trick again,

$$\begin{aligned} 0 &= \|x - w\|^2 + \|x^* - w^*\|^2 + 2\langle x - w, x^* - w^*\rangle \\ &\geq \|x - w\|^2 + \|x^* - w^*\|^2 - 2\|x - w\|\|x^* - w^*\| \geq 0, \end{aligned}$$

from which the rest of (10.4.1) now follows easily.

($\Longleftarrow$) Since (10.4.1) implies that $\langle x-w, x^*-w^*\rangle < 0$, this is immediate from the definition of maximality. ▮

The *duality map* $J\colon E \mapsto 2^{E^*}$ is defined by:

$$x^* \in Jx \iff \langle x, x^*\rangle = \|x\|^2 = \|x^*\|^2.$$

Further, $-J\colon E \mapsto 2^{E^*}$ is defined by:

$$(-J)x := -Jx.$$

Lemma 10.5. *Let* $(x, x^*) \in E \times E^*$. *Then*

$$(x, x^*) \in G(-J) \iff \|x\|^2 + \|x^*\|^2 + 2\langle x, x^*\rangle = 0.$$

Proof. Exercise!. ▮

Theorem 10.6 is the "$-J$" criterion for maximality.

Theorem 10.6. *Let* E *be reflexive and* M *be a monotone subset of* $E \times E^*$. *Then*

$$M \text{ is maximal monotone} \iff M + G(-J) = E \times E^*.$$

Proof. From Lemma 10.5 and Theorem 10.3, M is maximal monotone $\iff$ for all $(w, w^*) \in E \times E^*$, there exists $(x, x^*) \in M$ such that $(x-w, x^*-w^*) \in G(-J)$. But this last is equivalent to $(w-x, w^*-x^*) \in G(-J)$, i.e.,

$$(w, w^*) \in (x, x^*) + G(-J). \;▮$$

There is an obvious one–to–one correspondence between multifunctions from E into 2^{E^*} and subsets of $E \times E^*$: if $S\colon E \mapsto 2^{E^*}$ then $G(S)$ is the corresponding subset of $E \times E^*$, while if $G \subset E \times E^*$ then the corresponding multifunction is defined by

$$Sx := \{x^* \in E^*\colon (x, x^*) \in G\}.$$

Further, monotone multifunctions correspond to monotone subsets and maximal monotone multifunctions correspond to maximal monotone subsets. For the rest of these notes, we shall use whichever notation is more convenient. Now, it was proved by Minty that *if* E *is a Hilbert space and* $S\colon E \mapsto 2^{E^*}$ *is monotone then* S *is maximal monotone* $\iff R(S+J) = E^*$. Rockafellar showed that Minty's result can be extended to the case where E is reflexive and J and J^{-1} are single–valued. Further, it was proved by Asplund that any reflexive Banach space can be renormed so that J and J^{-1} are single–valued. (Of course, renorming does not affect monotonicity or maximality). This renorming theorem is not easy. We shall show in Theorem 10.7 that the implication ($\Longrightarrow$), known as "Rockafellar's surjectivity theorem", is true even without the renorming.

Theorem 10.7. *Let E be reflexive and $S\colon E \mapsto 2^{E^*}$ be maximal monotone. Then*

$$R(S+J) = E^*.$$

Proof. Let $w^* \in E^*$. From Theorem 10.6,

$$(0, w^*) \in G(S) + G(-J).$$

Thus there exist $x \in E$, $x^* \in Sx$ and $y^* \in (-J)(-x)$ such that $x^* + y^* = w^*$. But then $y^* \in Jx$, hence

$$w^* = x^* + y^* \in Sx + Jx \subset R(S+J). \blacksquare$$

Remark 10.8. We outline the proof that if E is reflexive and J and J^{-1} are single–valued then

$$R(S+J) = E^* \quad \Longrightarrow \quad S \text{ is maximal monotone.} \tag{10.8.1}$$

Suppose that $(z, z^*) \in E \times E^*$ is monotonically related to $G(S)$. Since $R(S+J) = E^*$, we can choose $(s, s^*) \in G(S)$ so that

$$s^* + Js = z^* + Jz. \tag{10.8.2}$$

(Remember that, for all $x \in E$, Jx is now a *point*.) We will show that

$$(z, z^*) = (s, s^*) \in G(S), \tag{10.8.3}$$

which will establish that S is maximal monotone. We have

$$\langle s - z, s^* - z^* \rangle + \langle s - z, Js - Jz \rangle = \langle s - z, (s^* + Js) - (z^* + Jz) \rangle = 0.$$

Since both terms on the left-hand side of the above equation are positive, they are both zero. In particular, $\langle s - z, Js - Jz \rangle = 0$, from which

$$\langle s, Js \rangle - \langle s, Jz \rangle - \langle z, Js \rangle + \langle z, Jz \rangle = 0, \tag{10.8.4}$$

which implies in turn that $\|s\|^2 - 2\|s\|\|z\| + \|z\|^2 \le 0$. It follows from this that $\|z\| = \|s\|$, hence $\|Jz\| = \|Js\| = \|z\| = \|s\|$. Substituting this in (10.8.4),

$$2\|s\|\|z\| - \langle s, Jz \rangle - \langle z, Js \rangle = 0,$$

hence $\langle z, Js \rangle = \|z\|\|Js\| = \|z\|^2 = \|Js\|^2$, that is to say

$$Js = Jz. \tag{10.8.5}$$

Substituting this in (10.8.2), $s^* = z^*$. Since J^{-1} is single–valued, we also obtain from (10.8.5) that $s = z$. Thus we have established (10.8.3), as required.

It was pointed out by S. Fitzpatrick (personal communication) that *if J is not single–valued then (10.8.1) does not follow*. Here is his reasoning: if J is not single–valued then there exist $x \in E$ and distinct elements y^* and z^* of Jx. Let $x^* := (y^* + z^*)/2$, and define S by setting $G(S) := G(J) \setminus \{(x, x^*)\}$. Now let $w^* \in E^*$. Since E is reflexive, there exists $w \in E$ such that $w^*/2 \in Jw$. If $(w, w^*/2) \neq (x, x^*)$ then $w^*/2 \in Sw$ hence

$$w^* = \frac{w^*}{2} + \frac{w^*}{2} \in Sw + Jw = (S + J)w.$$

If, on the other hand, $(w, w^*/2) = (x, x^*)$ then $w^* = 2x^* = y^* + z^*$. Since $y^* \in Jx$ and $y^* \neq x^*$, $y^* \in Sx$. Consequently,

$$w^* = y^* + z^* \in Sx + Jx = (S + J)x.$$

Thus we have proved that

$$w^* \in E^* \quad \Longrightarrow \quad w^* \in R(S + J),$$

that is to say, $R(S+J) = E^*$. On the other hand, S is obviously not maximal monotone.

It was pointed out by H. Bauschke (personal communication) that *if J is single–valued and J^{-1} is not single–valued then, again,* (10.8.1) *does not follow*. Here is his reasoning: there exist $z^* \in E^*$ and distinct elements x and y of $J^{-1}z^*$, and define S by setting $G(S) := (E \setminus \{x\}) \times \{0\}$. Now let $w^* \in E^*$. If $w^* = z^*$ then

$$w^* = 0 + z^* \in (S + J)y.$$

If, on the other hand, $w^* \neq z^*$ then, since E is reflexive, there exists $w \in E$ such that $w^* \in Jw$. Since J is single–valued, $z^* \in Jx$ and $w^* \neq z^*$, $w^* \notin Jx$. It follows that $w \neq x$, and so

$$w^* = 0 + w^* \in (S + J)w.$$

Thus we have proved that

$$w^* \in E^* \quad \Longrightarrow \quad w^* \in R(S + J),$$

that is to say, $R(S + J) = E^*$. On the other hand, S is, again, obviously not maximal monotone.

The following numerical estimate is sometimes useful:

Lemma 10.9. *Let E be reflexive, $\emptyset \neq G \subset E \times E^*$ and (x, x^*) satisfy (10.1.2). Then*

$$(s, s^*) \in G \quad \Longrightarrow \quad \|x\|^2 + \|x^*\|^2 \le 6\big(\|s\|^2 + \|s^*\|^2\big). \tag{10.9.1}$$

Proof. The proof of Lemma 10.1 shows that (x, x^*) satisfies (10.1.4) and, consequently,

$$\begin{aligned}(s, s^*) \in G \quad \Longrightarrow \quad & \|x\|^2 - 2\|x\|\|s^*\| + \|x^*\|^2 - 2\|x^*\|\|s\| \\ & \qquad\qquad \le 2\|s\|\|s^*\| \le \|s\|^2 + \|s^*\|^2 \\ \Longrightarrow \quad & (\|x\| - \|s^*\|)^2 + (\|x^*\| - \|s\|)^2 \le 2(\|s\|^2 + \|s^*\|^2).\end{aligned}$$

It follows from the triangle inequality in $\mathbb{R}^2$ that

$$\sqrt{\|x\|^2 + \|x^*\|^2} \le (1 + \sqrt{2})\sqrt{\|s\|^2 + \|s^*\|^2},$$

which gives (10.9.1). ▮

The result above leads to the following problem:

Problem 10.10. Find the smallest constant C such that if E is a reflexive Banach space and M is a maximal monotone subset of $E \times E^*$ then, for all $(x, x^*) \in M$ such that

$$\|x\|^2 + \|x^*\|^2 + 2\langle x, x^*\rangle = 0,$$

we must have

$$(s, s^*) \in M \quad \Longrightarrow \quad \|x\|^2 + \|x^*\|^2 \le C\big(\|s\|^2 + \|s^*\|^2\big).$$

We have from the proof of Lemma 10.9 that $C \le (1 + \sqrt{2})^2$.

11. Monotone multifunctions with bounded range

In the final section of this chapter, we shall use the minimax technique to prove Lemma 11.1, an extremely useful result. This result can also be established using the Debrunner–Flor extension theorem, which is outside the scope of these notes since its proof depends on Brouwer's fixed–point theorem (see Phelps, [35], Lemma 1.7, p. 4 and the comments preceding). Lemma 11.1 can also be established using the Farkas Lemma (see Fitzpatrick–Phelps, [23], Lemma 2.4, p. 580–581).

Lemma 11.1. *Let S: $E \mapsto 2^{E^*}$ be monotone and $R(S)$ be bounded.*
(a) Let $x \in E$. Then there exists $x^* \in E^*$ such that (x, x^*) is monotonically related to $G(S)$.
(b) If, in addition, S is maximal monotone then $D(S) = E$.

Proof. Define h: $\mathbb{R}^{(E\times E^*)} \times E^* \mapsto \mathbb{R}$ by

$$h(\mu, x^*) := r(\mu) - \langle p(\mu), x^* \rangle - \langle x, q(\mu) \rangle + \langle x, x^* \rangle. \tag{11.1.1}$$

Choose $K \geq 0$ such that

$$x^* \in R(S) \quad \Longrightarrow \quad \|x^*\| \leq K,$$

and let $A := \mathcal{CO}(S)$ and $B := \{x^* \in E^*\colon \|x^*\| \leq K\}$, with the topology $w(E^*, E)$. From the Banach–Alaoglu theorem, Theorem 4.1, B is compact. If $\mu \in A$ then, from the pqr–lemma, Lemma 9.1,

$$h(\mu, q(\mu)) \geq \langle p(\mu), q(\mu) \rangle - \langle p(\mu), q(\mu) \rangle - \langle x, q(\mu) \rangle + \langle x, q(\mu) \rangle = 0,$$

since $q(\mu) \in B$, it follows that

$$\inf_A \max_B h \geq 0.$$

The sets A and B are convex. Since h is affine in each variable and continuous on B, from the minimax theorem, Theorem 3.1,

$$\max_B \inf_A h \geq 0,$$

hence there exists $x^* \in B \subset E^*$ such that,

$$\mu \in A \quad \Longrightarrow \quad r(\mu) - \langle p(\mu), x^* \rangle - \langle x, q(\mu) \rangle + \langle x, x^* \rangle \geq 0.$$

Allowing μ to run through the values $\delta_{(s,s^*)}$ $((s, s^*) \in G(S))$, it follows that

$$(s, s^*) \in G(S) \quad \Longrightarrow \quad \langle s, s^* \rangle - \langle s, x^* \rangle - \langle x, s^* \rangle + \langle x, s^* \rangle \geq 0.$$

This completes the proof of (a), and (b) is an immediate consequence. ▌

Remark 11.2. It was pointed out by H. Bauschke (personal communication) that the result "dual" to Lemma 11.1 fails. Suppose that E is not reflexive. Let C be the unit ball of E and $S := N_C$ (see Section 8). From James's theorem, Theorem 4.11, $R(S) \neq E^*$. On the other hand, $D(S) = C$ is bounded.

III. A digression into convex analysis

12. Surrounding sets and the dom lemma

In this and the next section, we collect together some results on convex lower semicontinuous functions that we shall need for our later work. In this section, we give the "dom lemma", Lemma 12.2, which is a "quantitative" result, and the "dom corollary", Corollary 12.3, which is a "qualitative" result. Both these results will have their uses, the dom lemma in Lemma 17.2 and the dom corollary in Lemma 18.2 and Theorem 18.4. Both the dom lemma and the dom corollary are subsumed by the results of the next next section — we have treated them independently for essentially pedagogical reasons.

Let $x \in E$ and $A \subset E$. A is said to be *absorbing* if $\bigcup_{\lambda>0} \lambda A = E$. Any neighborhood of 0 is absorbing (exercise!). We write "$x \in \operatorname{sur} A$" and say that "$A$ surrounds x" if, for each $w \in E\backslash\{0\}$, there exists $\delta > 0$ such that $x+\delta w \in A$. The statement "$x \in \operatorname{sur} A$" is related to x being an "absorbing point" of A (see Phelps, [34], Definition 2.27(b), p. 28), but differs in that we do not require that $x \in A$. We also note that, if A is convex then $\operatorname{sur} A \subset A$ and so $\operatorname{sur} A$ is identical with the "core" or algebraic interior of A. In particular:

$$\text{if } A \text{ is convex then} \qquad (0 \in \operatorname{sur} A \iff A \text{ is absorbing}). \tag{12.0.1}$$

We start off with a classical result.

Lemma 12.1. *Let C be a closed convex absorbing set in E. Then C is a neighborhood of* 0.

Proof. Let $D := C \cap -C$. Then D is closed, convex and absorbing (exercise!) and $D = -D$, i.e., D is a "barrel". The result follows by applying Kelly–Namioka, [28], p. 104 to D.

Lemma 12.2. *Let $f \in \mathcal{PCLSC}(E)$ and $\operatorname{dom} f$ surround* 0. *Then there exist $\eta > 0$ and $n \geq 1$ such that*

$$w \in E \text{ and } \|w\| \leq \eta \quad \Longrightarrow \quad f(w) \leq n. \tag{12.2.1}$$

Proof. From (12.0.1),

$$\operatorname{dom} f \quad \text{is absorbing.} \tag{12.2.2}$$

In particular, $0 \in \operatorname{dom} f$. Let $n \geq f(0) \vee 0 + 1$, and put

$$C := \{w \in E\colon\ f(w) \leq n\}.$$

C is clearly closed and convex. We now show that C is absorbing. To this end, let w be an arbitrary element of E. From (12.2.2), there exist $\lambda > 0$ and $x \in \operatorname{dom} f$ such that $\lambda w = x$. We choose $\mu \in (0, 1]$ so that

$$\mu(f(x) - n + 1) \leq 1.$$

Then

$$\begin{aligned} f(\mu\lambda w) &= f(\mu x) \\ &\leq \mu f(x) + (1-\mu)f(0) \\ &\leq \mu f(x) + (1-\mu)(n-1) \\ &= \mu(f(x) - n + 1) + n - 1 \leq n. \end{aligned}$$

In other words, $\mu\lambda w \in C$. Thus C is absorbing, as required. It now follows from Lemma 12.1 that C is a neighborhood of 0, which gives the required result. ▌

Corollary 12.3. *Let $f \in \mathcal{PCLSC}(E)$. Then*

$$\operatorname{sur}(\operatorname{dom} f) = \operatorname{int}(\operatorname{dom} f).$$

Proof. Exercise!.

Remark 12.4. The classical "uniform boundedness theorem" can easily be deduced from the dom lemma. Here are the details: *Let F be a normed space and $\mathcal{B}$ be a nonempty pointwise bounded set of continuous linear operators from E into F. Then $\mathcal{B}$ is bounded in norm.*

Proof. Define $f\colon E \mapsto \mathbb{R}$ by

$$f(x) := \sup_{T \in \mathcal{B}} \|Tx\|.$$

Since $\operatorname{dom} f = E$, we can apply the dom lemma. It then follows from (12.2.1) that

$$T \in \mathcal{B} \quad \Longrightarrow \quad \|T\| \leq \frac{n}{\eta}.$$ ▌

The proof of the uniform boundedness theorem given above can be found in Holmes, [27], §17, p. 134. Lemma 12.2 also implies the result that a convex lower semicontinuous function is locally bounded on the interior of its domain. (See, for instance, Phelps, [34], Proposition 3.3, p. 39.)

13. The dom–dom lemma

We now come to the "dom–dom lemma", Lemma 13.1, which will be crucial for our analysis of the sums of maximal monotone operators in reflexive spaces. The dom–dom lemma is a "quantitative" result that also has a "qualitative" version, the "dom–dom corollary", Corollary 13.2. Both of these results will have their uses, the dom–dom lemma in Lemmas 14.1 and 21.1, and the dom–dom corollary in Theorems 22.1 and 22.2. The dom–dom lemma, which generalizes the open mapping theorem (see Remark 13.3) can itself be generalized considerably. (In this connection, we refer the reader to Robinson [39], Ursescu [58] and Borwein [9]). Here we confine our attention to what we will need in these notes. The idea for the proof of Lemma 13.1 is taken from Aubin–Ekeland, [3], Lemma 3.3.9, p. 136.

The dom lemma is an immediate consequence of the dom–dom lemma — put $g_1 := f$ and g_2 the "indicator function" of $\{0\}$. Consequently, the results of Sections 17 and 18 are much "simpler" than those of Sections 21–23.

We shall use the notation $\{E|\ g \le n,\ \|\ \| \le n\}$ as a shorthand for

$$\{x \in E\colon\ g(x) \le n,\ \|x\| \le n\}.$$

Lemma 13.1. *Let* $g_1,\ g_2 \in \mathcal{PCLSC}(E)$ *and* $\operatorname{dom} g_1 - \operatorname{dom} g_2$ *surround* 0. *Then there exists* $n \ge 1$ *such that*

$$\left.\begin{array}{r}\{E|\ g_1 \le n,\ \|\ \| \le n\} - \{E|\ g_2 \le n,\ \|\ \| \le n\}\\ \textit{is a neighborhood of } 0.\end{array}\right\} \tag{13.1.1}$$

Proof. From (12.0.1),

$$\operatorname{dom} g_1 - \operatorname{dom} g_2 \quad \text{is absorbing.} \tag{13.1.2}$$

In particular, $0 \in \operatorname{dom} g_1 - \operatorname{dom} g_2$, and so there exists $y \in \operatorname{dom} g_1 \cap \operatorname{dom} g_2$. Let $n \ge g_1(y) \vee g_2(y) \vee \|y\| + 1$, and put

$$X_1 := \{E|\ g_1 \le n,\ \|\ \| \le n\}, \quad X_2 := \{E|\ g_2 \le n,\ \|\ \| \le n\}$$

and

$$Z := X_1 - X_2.$$

Z is clearly bounded and convex. We first show that Z is absorbing. So let w be an arbitrary element of E. From (13.1.2), there exist $\lambda > 0$, $x_1 \in \operatorname{dom} g_1$ and $x_2 \in \operatorname{dom} g_2$ such that $\lambda w = x_1 - x_2$. We choose $\mu \in (0, 1]$ so that

$$\mu(g_1(x_1) \vee g_2(x_2) \vee \|x_1\| \vee \|x_2\| - n + 1) \le 1.$$

Then, for $i = 1,\ 2$,

$$\begin{aligned} g_i(\mu x_i + (1-\mu)y) &\le \mu g_i(x_i) + (1-\mu) g_i(y) \\ &\le \mu g_i(x_i) + (1-\mu)(n-1) \\ &= \mu(g_i(x_i) - n + 1) + n - 1 \le n \end{aligned}$$

and

$$\begin{aligned} \|\mu x_i + (1-\mu)y\| &\le \mu\|x_i\| + (1-\mu)\|y\| \\ &\le \mu\|x_i\| + (1-\mu)(n-1) \\ &= \mu(\|x_i\| - n + 1) + n - 1 \le n. \end{aligned}$$

Thus $\mu x_i + (1-\mu)y \in X_i$ and so

$$\mu\lambda w = \mu(x_1 - x_2) = [\mu x_1 + (1-\mu)y] - [\mu x_2 + (1-\mu)y] \in X_1 - X_2 = Z.$$

Since we can carry out this construction for all $w \in E$, this establishes that Z is absorbing, as required. Since $\overline{Z}$ is closed, convex and absorbing, it follows from Lemma 12.1 that $\overline{Z}$ is a neighborhood of 0 in E. Choose $\eta > 0$ so that

$$w \in E \text{ and } \|w\| \le 2\eta \quad \Longrightarrow \quad w \in \overline{Z}. \tag{13.1.3}$$

We shall prove that

$$w \in E \text{ and } \|w\| \le \eta \quad \Longrightarrow \quad w \in Z, \tag{13.1.4}$$

which will give the desired result. So let $w \in E$ and $\|w\| \le \eta$. Then, from (13.1.3), $2w \in \overline{Z}$, consequently

$$\text{there exists } z_1 \in Z \quad \text{such that} \quad \|2w - z_1\| \le \eta.$$

From (13.1.3) again, $4w - 2z_1 = 2(2w - z_1) \in \overline{Z}$, thus

$$\text{there exists } z_2 \in Z \quad \text{such that} \quad \|4w - 2z_1 - z_2\| \le \eta.$$

Continuing this argument, we find $z_1,\ z_2,\ z_3,\ \ldots \in Z$ such that, for all $k \ge 1$,

$$\|2^k w - 2^{k-1} z_1 - \cdots - z_k\| \le \eta,$$

from which

$$\|w - 2^{-1} z_1 - \cdots - 2^{-k} z_k\| \le 2^{-k}\eta,$$

hence

$$\sum_{k=1}^{\infty} 2^{-k} z_k = w.$$

For all $n \ge 1$, choose $x_{1,n} \in X_1$ and $x_{2,n} \in X_2$ such that $x_{1,n} - x_{2,n} = z_n$. Since X_1 and X_2 are bounded, closed and convex, there exist $v_1 \in X_1$ and $v_2 \in X_2$ such that

$$\sum_{k=1}^{\infty} 2^{-k} x_{1,n} = v_1 \quad \text{and} \quad \sum_{k=1}^{\infty} 2^{-k} x_{2,n} = v_2.$$

Then

$$\begin{aligned} v_1 - v_2 &= \sum_{k=1}^{\infty} 2^{-k} x_{1,n} - \sum_{k=1}^{\infty} 2^{-k} x_{2,n} \\ &= \sum_{k=1}^{\infty} 2^{-k} (x_{1,n} - x_{2,n}) \\ &= \sum_{k=1}^{\infty} 2^{-k} z_n \\ &= w, \end{aligned}$$

and so $w \in Z$. This completes the proof of (13.1.4), and hence also that of Lemma 13.1. ▌

Corollary 13.2. *Let $g_1,\ g_2 \in \mathcal{PCLSC}(E)$. Then*

$$\operatorname{sur}(\operatorname{dom} g_1 - \operatorname{dom} g_2) = \operatorname{int}(\operatorname{dom} g_1 - \operatorname{dom} g_2).$$

Proof. We shall prove that

$$\operatorname{sur}(\operatorname{dom} g_1 - \operatorname{dom} g_2) \subset \operatorname{int}(\operatorname{dom} g_1 - \operatorname{dom} g_2). \tag{13.2.1}$$

This gives the desired result, since the reverse inclusion is trivial. So let

$$x \in \operatorname{sur}(\operatorname{dom} g_1 - \operatorname{dom} g_2).$$

Define $g_3\colon E \mapsto \mathbb{R} \cup \{\infty\}$ by $g_3(y) := g_1(y+x) \quad (y \in E)$. Then $\operatorname{dom} g_3 = \operatorname{dom} g_1 - x$, from which

$$0 \in \operatorname{sur}(\operatorname{dom} g_3 - \operatorname{dom} g_2).$$

From Lemma 13.1,

$$0 \in \operatorname{int}(\operatorname{dom} g_3 - \operatorname{dom} g_2) = \operatorname{int}(\operatorname{dom} g_1 - x - \operatorname{dom} g_2).$$

Thus $x \in \operatorname{int}(\operatorname{dom} g_1 - \operatorname{dom} g_2)$, which completes the proof of (13.2.1). ▌

Remark 13.3. The classical "open mapping theorem" can easily be deduced from the dom–dom lemma. Here are the details. We first observe by applying the dom–dom lemma to indicator functions that *if C_1 and C_2 are closed convex subsets of E and $C_1 - C_2$ surrounds 0 then there exists $n \geq 1$ such that*

$$\{x \in C_1\colon \|x\| \leq n\} - \{x \in C_2\colon \|x\| \leq n\} \quad \text{is a neighborhood of } 0.$$

If now F and H are Banach spaces and $T \in B(F, H)$ is surjective then, for all $(x, y) \in F \times H$, there exists $z \in F$ such that that $y = Tz$, and consequently

$$(x, y) = (x, Tz) = (z, Tz) - (z - x, 0) \in G(T) - (F \times \{0\}).$$

Applying the above result with $E := F \times H$ normed by

$$\|(x, y)\| := \sqrt{\|x\|^2 + \|y\|^2},$$

$C_1 := G(T)$ and $C_2 := F \times \{0\}$, there exist $n \geq 1$ and $\eta > 0$ such that

$$\begin{aligned} w \in H \text{ and } \|w\| \leq \eta \quad &\Longrightarrow \\ &(0, w) \in \{(z, Tz)\colon\ z \in F,\ \|z\| \leq n\} - (F \times \{0\}) \\ &\Longrightarrow \quad w \in \{Tz\colon\ z \in F,\ \|z\| \leq n\}. \end{aligned}$$

It follows that T is an open mapping. ∎

Thus the dom–dom lemma is both a generalization of the open mapping theorem and, in some sense, a "second order" generalization of the uniform boundedness theorem.

Remark 13.4. As we have observed, Lemma 13.1 is a generalization of Lemma 12.2. In this remark, we shall sketch a generalization of Lemma 12.2 in a totally different direction.
(a) Let B be a nonmeager Borel set in E (that is, a Borel set of the second category). Then $B - B$ is a neighborhood of 0.
(b) Let D be a convex absorbing Borel set in E and D be symmetric, i.e., $D = -D$. Then D is a neighborhood of 0.
(c) Let C be a convex absorbing Borel set in E. Then C is a neighborhood of 0.
(d) Let C be a convex Borel set in E. Then $\operatorname{sur} C = \operatorname{int} C$.
(e) Let $f\colon E \mapsto \mathbb{R} \cup \{\infty\}$ be a convex Borel function and $\operatorname{dom} f$ surround 0. *Then there exist $\eta > 0$ and $n \geq 1$ such that*

$$w \in E \text{ and } \|w\| \leq \eta \quad \Longrightarrow \quad f(w) \leq n.$$

Proof. (a) Any Borel set satisfies the "condition of Baire", that is to say, there exists an open set U such that $U \setminus B$ and $B \setminus U$ are meager, and so (a) follows from the "difference theorem". See Kelly–Namioka, [28], 10.4, p. 92 and the discussion preceding.

(b) It follows from Baire's theorem that E, being a complete metric space, is nonmeager. Since $\bigcup_{n \geq 1} nD = E$ there exists $n \geq 1$ such that nD is nonmeager, from which $(1/2)D$ is nonmeager. Since D is convex and symmetric,

$$D = \frac{1}{2}D + \frac{1}{2}D = \frac{1}{2}D - \frac{1}{2}D,$$

thus it follows from (a) that D is a neighborhood of 0.

(c) Let $D := C \cap -C$. Then D is a convex absorbing Borel set and $D = -D$. From (b), D is a neighborhood of 0, from which C is a neighborhood of 0 also.

(d) is immediate from (c), a translation argument and (12.0.1).

(e) From (12.0.1), $\operatorname{dom} f$ is absorbing. In particular, $0 \in \operatorname{dom} f$. Let $n \geq f(0) \vee 0 + 1$, and put $C := \{x \in E\colon f(x) \leq n\}$. C is clearly convex. We first show that C is absorbing. So let w be an arbitrary element of E. Since $\operatorname{dom} f$ is absorbing, there exist $\lambda > 0$, and $x \in \operatorname{dom} f$ such that $\lambda w = x$. We choose $\mu \in (0, 1]$ so that $\mu(f(x) - n + 1) \leq 1$. Then

$$\begin{aligned} f(\mu\lambda w) = f(\mu x) = f(\mu x + (1-\mu)0) &\leq \mu f(x) + (1-\mu) f(0) \\ &\leq \mu f(x) + (1-\mu)(n-1) \\ &= \mu(f(x) - n + 1) + n - 1 \leq n \end{aligned}$$

Thus $\mu\lambda w \in C$. Since we can carry out this construction for all $w \in E$, this establishes that C is absorbing, as required. It now follows from (c) that C is a neighborhood of 0.

Remark 13.5. Lemma 13.1 and Remark 13.4 suggest the following question:

Problem 13.6. Let $g_1,\ g_2\colon E \mapsto \mathrm{IR} \cup \{\infty\}$ be convex Borel functions and $\operatorname{dom} g_1 - \operatorname{dom} g_2$ surround 0. Does there necessarily exist $n \geq 1$ such that

$$\{E|\ g_1 \leq n,\ \|\ \| \leq n\} - \{E|\ g_2 \leq n,\ \|\ \| \leq n\} \quad \text{is a neighborhood of 0?}$$

In particular: Let B_1 and B_2 be convex Borel sets in E and $B_1 - B_2$ be absorbing. Is $B_1 - B_2$ necessarily a neighborhood of 0?

14. The dom–dom lemma and the Attouch–Brézis condition

Our main motivation for proving the dom–dom lemma, Lemma 13.1, is as an intermediate step towards the analysis in Sections 21–23 of "constraint qualifications" for the maximal monotonicity of the sum of maximal monotone multifunctions on a reflexive space. However, we pause here to give another application of the dom–dom lemma, which leads to the Attouch–Brézis version of the Fenchel duality theorem, Theorem 14.2.

Lemma 14.1. *Let $f_1,\ f_2 \in \mathcal{PCLSC}(E)$,*

$$F := \bigcup_{\lambda > 0} \lambda(\operatorname{dom} f_1 - \operatorname{dom} f_2) \quad \textit{be a closed subspace of } E$$

and

$$f_1 + f_2 \geq 0 \textit{ on } E.$$

Then there exists $M > 0$ such that

$$\text{for all } x_1,\ x_2 \in E, \quad f_1(x_1) + f_2(x_2) + M\|x_1 - x_2\| \geq 0. \tag{14.1.1}$$

Proof. Since $0 \in F$, there exists $z \in \operatorname{dom} f_1 \cap \operatorname{dom} f_2$. Define $g_1,\ g_2\colon E \mapsto \mathbb{R} \cup \{\infty\}$ by $g_1(x) := f_1(x+z)$ and $g_2(x) := f_2(x+z)$ $(x \in E)$. Then $\operatorname{dom} g_1 \subset F$, $\operatorname{dom} g_2 \subset F$ and $\operatorname{dom} g_1 - \operatorname{dom} g_2$ surrounds 0 in F. From the dom–dom lemma, Lemma 13.1, (with E replaced by F) there exist $n \geq 1$ and $\eta > 0$ such that

$$\left.\begin{aligned} &w \in F \text{ and } \|w\| \leq \eta \quad \Longrightarrow \\ &\qquad w \in \{x \in F\colon\ g_1(x) \leq n\} - \{x \in F\colon\ g_2(x) \leq n\}. \end{aligned}\right\} \tag{14.1.2}$$

We put $M := 2n/\eta$, and we shall prove that (14.1.1) is satisfied. Since $f_1 + f_2 \geq 0$ on E, (14.1.1) is immediate if $x_1 = x_2$, so we can and will assume that $x_1 \neq x_2$. Let

$$\lambda := \frac{\eta}{\|x_2 - x_1\|} > 0$$

and

$$w := \lambda(x_2 - x_1).$$

Now (14.1.1) is also immediate if $x_1 \notin \operatorname{dom} f_1$ or $x_2 \notin \operatorname{dom} f_2$, so we can and will assume that $x_1 \in \operatorname{dom} f_1$ and $x_2 \in \operatorname{dom} f_2$. Thus $w \in -F = F$. Since $\|w\| = \eta$, it follows from (14.1.2) that there exist $y_1,\ y_2 \in F$ such that

$$g_1(y_1) \leq n, \ g_2(y_2) \leq n \quad \text{and} \quad y_1 - y_2 = w. \tag{14.1.3}$$

We derive from this that $y_1 + \lambda x_1 = y_2 + \lambda x_2$, hence, since $g_1 + g_2 \geq 0$ on E,

$$g_1\left(\frac{y_1 + \lambda(x_1 - z)}{1+\lambda}\right) + g_2\left(\frac{y_2 + \lambda(x_2 - z)}{1+\lambda}\right) \geq 0.$$

Thus, using the convexity of g_1 and g_2,

$$g_1(y_1) + \lambda g_1(x_1 - z) + g_2(y_2) + \lambda g_2(x_2 - z) \geq 0.$$

Combining this with (14.1.3), we derive that

$$\lambda\big[g_1(x_1 - z) + g_2(x_2 - z)\big] + 2n \geq 0.$$

Substituting in the value of $\lambda = \eta/\|x_1 - x_2\|$, we obtain

$$\frac{\eta\big[g_1(x_1 - z) + g_2(x_2 - z)\big]}{\|x_1 - x_2\|} + 2n \geq 0,$$

and (14.1.1) now follows from the definitions of g_1, g_2 and M. ∎

Theorem 14.2. *Let* $f_1,\ f_2 \in \mathcal{PCLSC}(E)$,

$$\bigcup_{\lambda>0} \lambda(\operatorname{dom} f_1 - \operatorname{dom} f_2) \quad \textit{be a closed subspace of } E$$

and

$$f_1 + f_2 \geq 0 \text{ on } E.$$

Then

$$\textit{there exists } z^* \in E^* \textit{ such that} \quad f_1^*(-z^*) + f_2^*(z^*) \leq 0.$$

Proof. This follows from Lemma 14.1 and the discussion in Section 6, since the conclusion of Lemma 14.1 is identical with (6.2.2). ∎

IV. General monotone multifunctions

15. Two convex functions determined by a multifunction

We now end our digression, and return to our analysis of multifunctions. We suppose throughout this section that $S\colon E \mapsto 2^{E^*}$ is nontrivial. We shall abuse our notation a little, and use $\mathcal{CO}(S)$ as an abbreviation for $\mathcal{CO}(G(S))$.

Definition 15.1. We define the function $\psi_S\colon E \mapsto \mathbb{R} \cup \{\infty\}$ by

$$\psi_S(x) := \sup_{(w,w^*)\in G(S)} \frac{\langle x - w, w^*\rangle}{1 + \|w\|} \quad (x \in E)$$

and the function $\chi_S\colon E \mapsto \mathbb{R} \cup \{\infty\}$ by

$$\chi_S(x) := \sup_{\mu \in \mathcal{CO}(S)} \frac{\langle x, q(\mu)\rangle - r(\mu)}{1 + \|p(\mu)\|} \quad (x \in E).$$

Since ψ_S and χ_S are the suprema of a family of continuous affine functions, they are both convex and lower semicontinuous. The function ψ_S was first defined in our paper [53], while the function χ_S was first defined, and many of its properties were established in the paper [18] by Coodey–Simons. Clearly, $\psi_S \le \chi_S$ on E.

We take time off to give an example showing that the functions ψ_S and χ_S do not always coincide. It will also indicate that it will be extremely difficult to compute them explicitly in all but the most trivial cases. Fortunately, for the applications we have in mind, it will not be necessary to compute them explicitly. Define $S\colon \mathbb{R} \mapsto 2^{\mathbb{R}}$ by

$$Sx := \begin{cases} \{0\}, & \text{if } x \le 0; \\ \{x\}, & \text{otherwise.} \end{cases}$$

Then S is maximal monotone and

$$\psi_S(x) = 0 \vee \sup_{w>0} \frac{(x - w)(w)}{1 + w} = \begin{cases} 0, & \text{if } x \le 0; \\ (\sqrt{x+1} - 1)^2, & \text{otherwise.} \end{cases}$$

On the other hand,

$$\begin{aligned}\chi_S(x) &\geq \sup_{w>0} \frac{x(\frac{0}{2}+\frac{2w}{2}) - (\frac{-2w\cdot 0}{2}+\frac{2w\cdot 2w}{2})}{1+|\frac{-2w}{2}+\frac{2w}{2}|}\\ &= \sup_{w>0}(xw-2w^2)\\ &= \begin{cases} 0, & \text{if } x \leq 0;\\ x^2/8, & \text{otherwise.}\end{cases}\end{aligned}$$

It follows from this that

$$\text{for all } x \geq 3, \quad \psi_S(x) < \chi_S(x).$$

(Exercise!) This is a simplified version of a result proved by Coodey in [17], Example 3.7, p. 41–46 that *if* $S\colon \mathbb{R} \mapsto 2^{\mathbb{R}}$ *is defined by* $S(x) := \{e^x\}$ *then,*

$$\textit{for all } x > 1, \quad \psi_S(x) < \chi_S(x).$$

Why do we define *two* convex functions on E determined by S? Since ψ_S has a simpler definition than χ_S, we prefer to prove results about ψ_S whenever we can. However, χ_S has the advantage that the set "indexing" it is convex, which give us the possibility of the minimax technique as a tool. Indeed, there are many cases where we know that a result using χ_S is true, but we do not know whether the same holds for the analogous result with χ_S replaced by ψ_S. The "D–dom lemma", Lemma 15.2 (which we shall use a number of times in the sequel) shows that if S is monotone then $\operatorname{dom}\chi_S \neq \emptyset$. We use the notation "co" to stand for "convex hull of".

Lemma 15.2. *Let* $S\colon E \mapsto 2^{E^*}$ *be nontrivial and monotone. Then*

$$D(S) \subset \operatorname{co} D(S) \subset \operatorname{dom}\chi_S \subset \operatorname{dom}\psi_S.$$

Proof. The last inclusion follows from the fact that $\psi_S \leq \chi_S$ on E. Since $\operatorname{dom}\chi_S$ is convex, it remains to prove that

$$D(S) \subset \operatorname{dom}\chi_S. \tag{15.2.1}$$

To this end, let $w \in D(S)$. Pick $w^* \in Sw$, and define $\beta := \langle w, w^*\rangle \vee \|w^*\|$. If $(s, s^*) \in G(S)$ then, since S is monotone,

$$\langle w, s^*\rangle - \langle s, s^*\rangle = \langle w - s, s^*\rangle \leq \langle w - s, w^*\rangle = \langle w, w^*\rangle - \langle s, w^*\rangle.$$

Thus, for all $\mu \in \mathcal{CO}(S)$ and $(s, s^*) \in G(S)$,

$$\langle w, \mu_{(s,s^*)}s^*\rangle - \mu_{(s,s^*)}\langle s, s^*\rangle \leq \mu_{(s,s^*)}\langle w, w^*\rangle - \langle \mu_{(s,s^*)}s, w^*\rangle.$$

Summing over $(s, s^*) \in G(S)$,

$$\begin{aligned}\langle w, q(\mu)\rangle - r(\mu) &\leq \langle w, w^*\rangle - \langle p(\mu), w^*\rangle\\ &\leq \langle w, w^*\rangle + \|p(\mu)\|\|w^*\|\\ &\leq \beta(1 + \|p(\mu)\|).\end{aligned}$$

Dividing by $1 + \|p(\mu)\|$, and taking the supremum over $\mu \in \mathcal{CO}(S)$, we see that $\chi_S(w) \leq \beta$, which implies that $w \in \operatorname{dom}\chi_S$. This completes the proof of (15.2.1), and hence that of Lemma 15.2. ▌

Lemma 15.3. *Let $T\colon E \mapsto 2^{E^*}$ be nontrivial and $w \in E$. Let $S := (T^{-1} - w)^{-1}$. Then $\operatorname{dom}\chi_S = \operatorname{dom}\chi_T - w$ and $\operatorname{dom}\psi_S = \operatorname{dom}\psi_T - w$.*

Proof. We give the details for χ_S, the details for ψ_S are similar but simpler. We shall prove that, for all $x \in E$,

$$\chi_S(x) \le (1+\|w\|)|\chi_T(x+w)| \quad \text{and} \quad \chi_T(x+w) \le (1+\|w\|)|\chi_S(x)|, \tag{15.3.1}$$

from which the desired result follows. It fact, it suffices to prove the second inequality in (15.3.1) — the first inequality follows by replacing w by $-w$ and interchanging the roles of T and S. Let $\mu \in \mathcal{CO}(T)$, and define $\nu \in \mathcal{CO}(S)$ by $\nu_{(s,s^*)} = \mu_{(s+w,s^*)}$. We note that

$$p(\nu) = p(\mu) - w, \quad q(\mu) = q(\nu) \quad \text{and} \quad r(\mu) - \langle w, q(\mu)\rangle = r(\nu).$$

Thus

$$\langle x + w, q(\mu)\rangle - r(\mu) = \langle x, q(\nu)\rangle - r(\nu)$$

from the definition of $\chi_S(x)$,

$$\begin{aligned} &\le (1 + \|p(\nu)\|)\chi_S(x) \\ &= (1 + \|p(\mu) - w\|)\chi_S(x) \\ &\le (1 + \|p(\mu)\| + \|w\|)|\chi_S(x)| \\ &\le (1 + \|p(\mu)\|)(1 + \|w\|)|\chi_S(x)|. \end{aligned}$$

The second inequality in (15.3.1) follows by dividing by $1 + \|p(\mu)\|$, taking the supremum over μ, and using the definition of $\chi_T(x + w)$. This completes the proof of Lemma 15.3. ∎

Lemma 15.4. *Let $S\colon E \mapsto 2^{E^*}$ be nontrivial and $w^* \in E^*$. Let $T := S - w^*$. Then $\operatorname{dom}\chi_T = \operatorname{dom}\chi_S$ and $\operatorname{dom}\psi_T = \operatorname{dom}\psi_S$.*

Proof. We give the details for χ_S, the details for ψ_S are similar but simpler. Let $x \in E$ and $\beta := \|w^*\| \vee \|x\|\|w^*\|$. We shall prove that

$$\chi_T(x) - \beta \le \chi_S(x) \le \chi_T(x) + \beta, \tag{15.4.1}$$

from which the desired result follows. It fact, it suffices to prove the second inequality in (15.4.1) — the first inequality follows by replacing w^* by $-w^*$ and interchanging the roles of S and T. Let $\mu \in \mathcal{CO}(S)$, and define $\nu \in \mathcal{CO}(T)$ by $\nu_{(t,t^*)} = \mu_{(t,t^*+w^*)}$. We note that

$$p(\nu) = p(\mu), \quad q(\mu) = q(\nu) + w^* \quad \text{and} \quad r(\mu) = r(\nu) + \langle p(\nu), w^*\rangle.$$

Thus

$$\begin{aligned}\langle x, q(\mu)\rangle - r(\mu) &= \langle x, q(\nu) + w^*\rangle - r(\nu) - \langle p(\nu), w^*\rangle \\ &\le \langle x, q(\nu)\rangle - r(\nu) + \|x\|\|w^*\| + \|p(\nu)\|\|w^*\| \\ &\le \langle x, q(\nu)\rangle - r(\nu) + \beta + \beta\|p(\nu)\| \\ &= \langle x, q(\nu)\rangle - r(\nu) + \beta(1 + \|p(\nu)\|)\end{aligned}$$

from the definition of $\chi_T(x)$,

$$\begin{aligned}&\le (1 + \|p(\nu)\|)\chi_T(x) + \beta(1 + \|p(\nu)\|) \\ &= (1 + \|p(\mu)\|)(\chi_T(x) + \beta).\end{aligned}$$

The result follows by dividing by $1 + \|p(\mu)\|$, taking the supremum over μ, and using the definition of $\chi_S(w)$. ∎

The following facts are known about ψ_S and χ_S and the relationship between them:

• *If S is maximal monotone then*

$$\psi_S \ge 0 \text{ on } E.$$

(See Coodey, [17], Lemma 3.1, p. 35–36.)

• *If $S\colon E \mapsto 2^{E^*}$ is nontrivial and $D(S)$ is bounded then*

$$\operatorname{dom}\psi_S = \operatorname{dom}\chi_S.$$

(See Coodey, [17], Theorem 3.8, p. 46–47.)

• *If $S\colon E \mapsto 2^{E^*}$ is maximal monotone and $D(S)$ is closed and convex then*

$$\operatorname{dom}\chi_S = \operatorname{dom}\psi_S = D(S).$$

(See Theorem 16.2.)

• *If E is reflexive and $S\colon E \mapsto 2^{E^*}$ is maximal monotone then*

$$\overline{\operatorname{dom}\chi_S} = \overline{\operatorname{dom}\psi_S} = \overline{D(S)}.$$

(See Theorem 18.6.)

• If E is not reflexive then it is a "hard problem" to find maximal monotone $S\colon E \mapsto 2^{E^*}$ such that

$$\overline{\operatorname{dom}\chi_S} \ne \overline{\operatorname{dom}\psi_S}.$$

(See the discussion preceding Lemma 26.2.)

• *If T is the subdifferential of the convex function $f\colon \mathbb{R} \mapsto \mathbb{R}$ defined by $f(x) := |x|$ and $S = -T$ then*

$$\operatorname{dom}\chi_S \textit{ is a proper subset of } \operatorname{dom}\psi_S.$$

(Exercise!)

Since S in the example above is "antimonotone", this suggests the following problem:

Problem 15.5. Find a Banach space E and a maximal monotone multifunction $S\colon\ E \mapsto 2^{E^*}$ such that

$$\operatorname{dom}\chi_S \quad \text{is a proper subset of} \quad \operatorname{dom}\psi_S.$$

16. Maximal monotonicity and closed convex sets

The main results of this section are:

• Theorem 16.2, in which we prove that *if $S\colon\ E \mapsto 2^{E^*}$ is maximal monotone and $D(S)$ is closed and convex then*

$$\operatorname{dom}\chi_S = \operatorname{dom}\psi_S = D(S),$$

• Theorem 16.8, in which we prove that *if $S_1\colon\ E \mapsto 2^{E^*}$ and $S_2\colon\ E \mapsto 2^{E^*}$ are maximal monotone then*

$$\overline{\operatorname{lin}(\operatorname{dom}\chi_{S_1} - \operatorname{dom}\chi_{S_2})} = \overline{\operatorname{lin}(\operatorname{dom}\psi_{S_1} - \operatorname{dom}\psi_{S_2})} = \overline{\operatorname{lin}(D(S_1) - D(S_2))}$$

and

$$\overline{\operatorname{aff}(\operatorname{dom}\chi_{S_1} - \operatorname{dom}\chi_{S_2})} = \overline{\operatorname{aff}(\operatorname{dom}\psi_{S_1} - \operatorname{dom}\psi_{S_2})} = \overline{\operatorname{aff}(D(S_1) - D(S_2))},$$

where "lin" stands for "linear span of" and "aff" stands for "affine hull of", and

• Theorem 16.10, in which we suppose that F is a closed subspace of E, $T\colon\ E \mapsto 2^{E^*}$ is monotone and $\emptyset \neq D(T) \subset F$ and prove that T *is maximal monotone if, and only if, ($F^\perp$ is defined in Definition 16.4)*

$$\text{for all } u \in D(T), \quad Tu + F^\perp = Tu.$$

and T_F is maximal monotone, where $T_F\colon\ F \mapsto 2^{F^*}$ is defined by

$$T_F x := \{x^*|_F\colon\ x^* \in Tx\}.$$

The material in this section is a generalization of results that appeared in [56] and [57].

In this section, we suppose that C is a closed convex subset of E, and we recall that the normality multifunction $N_C\colon\ E \mapsto 2^{E^*}$ is defined by

$$(x, x^*) \in G(N_C) \iff x \in C \text{ and } \langle x, x^*\rangle = \max_C x^*. \tag{8.1.1}$$

We first prove a simple lemma:

Lemma 16.1. *Let $S\colon E \mapsto 2^{E^*}$ be maximal monotone and $D(S) \subset C$. Then*

$$S + N_C = S. \tag{16.1.1}$$

Proof. If $x \in D(S) \subset C = D(N_C)$ then $0 \in N_C(x)$, so the inclusion "$\supset$" in (16.1.1) is clear. We now prove the opposite inclusion. Let $(x, x^*) \in G(S)$ and $(x, y^*) \in G(N_C)$. Since S is monotone,

$$\text{for all } (s, s^*) \in G(S), \quad \langle x - s, x^* - s^* \rangle \geq 0$$

and, from the definition of N_C,

$$\text{for all } (s, s^*) \in G(S), \quad \langle x - s, y^* \rangle \geq 0.$$

Adding the two inequalitites above,

$$\text{for all } (s, s^*) \in G(S), \quad \langle x - s, x^* + y^* - s^* \rangle \geq 0$$

thus, since S is maximal monotone, $(x, x^* + y^*) \in G(S)$. This establishes the inclusion "$\subset$" in (16.1.1), and completes the proof of the lemma. ▌

We now use Lemma 16.1 to give a precise description of $\operatorname{dom} \chi_S$ and $\operatorname{dom} \psi_S$ when S is maximal monotone and $D(S)$ is closed and convex.

Theorem 16.2. *Let $S\colon E \mapsto 2^{E^*}$ be maximal monotone and $D(S)$ be closed and convex. Then*

$$\operatorname{dom} \chi_S = \operatorname{dom} \psi_S = D(S).$$

Proof. We shall prove that

$$\operatorname{dom} \psi_S \subset D(S).$$

The required result will then follow from the D–dom lemma, Lemma 15.2. To this end, suppose that $x \in \operatorname{dom} \psi_S$. Let $C := D(S)$ and (y, y^*) be an arbitrary element of $G(N_C)$. Then $y \in D(S)$, so there exists $s^* \in Sy$. Let $\lambda \geq 0$. Then $\lambda y^* \in N_C(y)$ hence, using Lemma 16.1,

$$s^* + \lambda y^* \in (S + N_C)(y) = Sy.$$

Thus

$$\frac{\langle x - y, s^* + \lambda y^* \rangle}{1 + \|y\|} \leq \psi_S(x)$$

hence

$$\lambda \langle x - y, y^* \rangle \leq (1 + \|y\|)\psi_S(x) - \langle x - y, s^* \rangle.$$

Letting $\lambda \mapsto \infty$,

$$\langle x - y, y^* \rangle \leq 0,$$

i.e.,

$$\langle x - y, 0 - y^* \rangle \geq 0.$$

However, as we have already observed in Section 8, N_C is maximal monotone, Consequently, $(x, 0) \in G(N_C)$, that is to say, $x \in C = D(S)$, as required. ▌

We note from the discussion in Section 8 that the proof of Theorem 16.2 uses implicitly either Rockafellar's maximal monotonicity theorem or the Bishop–Phelps theorem. Theorem 16.2 might give one grounds to hope for a positive solution to the following problem:

Problem 16.3. Let $S\colon E \mapsto 2^{E^*}$ be maximal monotone. Then is it necessarily true that

$$\overline{\operatorname{dom}\chi_S} = \overline{\operatorname{co}D(S)}?$$

We shall see in Theorem 26.3 that the above *is* true if E is reflexive or, more generally, S is "of type (FPV)". We shall also see in Corollary 16.9 that we *do* always have:

$$\overline{\operatorname{aff}(\operatorname{dom}\chi_S)} = \overline{\operatorname{aff}(\operatorname{dom}\psi_S)} = \overline{\operatorname{aff}D(S)}.)$$

In preparation for this and other more general results, we first analyze how maximal monotonicity interacts with closed subspaces of E.

Definition 16.4. Let F be a subspace of E, and write

$$F^\perp := \{y^* \in E^*\colon \langle F, y^*\rangle = \{0\}\}.$$

Let $S\colon E \mapsto 2^{E^*}$. We say that S is F*-saturated* if

$$\text{for all } u \in D(S), \quad Su + F^\perp = Su.$$

Lemma 16.5. *Let F be a closed subspace of E, $w \in E$, $S\colon E \mapsto 2^{E^*}$ be maximal monotone and $D(S) \subset F + w$. Then S is F-saturated.*

Proof. This is immediate from Lemma 16.1 since, for all $u \in D(S) \subset F+w$, $N_{F+w}(u) = F^\perp$. ∎

The proof of the next lemma is similar to that of Theorem 16.2, except that it only uses a separation theorem rather than the Bishop–Phelps theorem or Rockafellar's maximal monotonicity theorem.

Lemma 16.6. *Let F be a closed subspace of E, $w \in E$, $S\colon E \mapsto 2^{E^*}$ be nontrivial and F-saturated, and $D(S) \subset F + w$. Then*

$$\operatorname{dom}\psi_S \subset F + w.$$

Proof. Let $x \in \operatorname{dom}\psi_S$ and $u \in D(S)$. Fix $u^* \in Su$. We first prove that

$$y^* \in F^\perp \implies \langle x - u, y^*\rangle = 0. \tag{16.6.1}$$

Let $y^* \in F^\perp$. Let λ be an arbitrary real number. Since S is F-saturated, $(u, u^* + \lambda y^*) \in G(S)$. Thus, from the definition of $\psi_S(x)$,

$$\frac{\langle x - u, u^* + \lambda y^*\rangle}{1 + \|u\|} \le \psi_S(x)$$

hence

$$\lambda\langle x - u, y^*\rangle \le (1 + \|u\|)\psi_S(x) - \langle x - u, u^*\rangle.$$

Since this holds for all $\lambda \in \mathbb{R}$, $\langle x - u, y^*\rangle = 0$, which completes the proof of (16.6.1). Now F is closed, thus it follows from (16.6.1) and Theorem 4.4 that $x - u \in F$. Since $u \in D(S) \subset F + w$,

$$x = (x - u) + u \in F + F + w = F + w. ∎$$

By combining Lemmas 16.5 and 16.6, we obtain the following very useful result:

Corollary 16.7. *Let F be a closed subspace of E, $w \in E$, $S\colon E \mapsto 2^{E^*}$ be maximal monotone and $D(S) \subset F + w$. Then*

$$\operatorname{dom} \psi_S \subset F + w.$$

In our analysis of constraint qualifications for pairs of maximal monotone multifunctions in Section 23, we will need to examine the closed subspaces $\overline{\operatorname{lin}(D(S_1) - D(S_2))}$ and $\overline{\operatorname{lin}(\operatorname{dom} \chi_{S_1} - \operatorname{dom} \chi_{S_2})}$. In this connection, the second part of the following result will be important:

Theorem 16.8. *Let $S_1\colon E \mapsto 2^{E^*}$ and $S_2\colon E \mapsto 2^{E^*}$ be maximal monotone. Then*

$$\overline{\operatorname{aff}(\operatorname{dom} \chi_{S_1} - \operatorname{dom} \chi_{S_2})} = \overline{\operatorname{aff}(\operatorname{dom} \psi_{S_1} - \operatorname{dom} \psi_{S_2})} = \overline{\operatorname{aff}(D(S_1) - D(S_2))}$$

and

$$\overline{\operatorname{lin}(\operatorname{dom} \chi_{S_1} - \operatorname{dom} \chi_{S_2})} = \overline{\operatorname{lin}(\operatorname{dom} \psi_{S_1} - \operatorname{dom} \psi_{S_2})} = \overline{\operatorname{lin}(D(S_1) - D(S_2))}.$$

Proof. We prove the "affine" case. The "linear" case follows by setting $z = 0$ in what follows. By virtue of the D–dom lemma, Lemma 15.2, we only have to prove that

$$\overline{\operatorname{aff}(\operatorname{dom} \psi_{S_1} - \operatorname{dom} \psi_{S_2})} \subset \overline{\operatorname{aff}(D(S_1) - D(S_2))},$$

for which it clearly suffices to prove that

$$\operatorname{dom} \psi_{S_1} - \operatorname{dom} \psi_{S_2} \subset \overline{\operatorname{aff}(D(S_1) - D(S_2))}. \tag{16.8.1}$$

Now there exist a closed subspace F of E and $z \in E$ such that

$$\overline{\operatorname{aff}(D(S_1) - D(S_2))} = F + z.$$

Let y be an arbitrary element of $D(S_2)$. Then

$$D(S_1) - y \subset D(S_1) - D(S_2) \subset F + z,$$

from which

$$D(S_1) \subset F + y + z.$$

Using Corollary 16.7, we deduce from this that

$$\operatorname{dom} \psi_{S_1} \subset F + y + z,$$

i.e.,

$$\operatorname{dom} \psi_{S_1} - y \subset F + z.$$

Allowing y to run through $D(S_2)$, we have proved that

$$\operatorname{dom} \psi_{S_1} - D(S_2) \subset F + z. \tag{16.8.2}$$

Let x be an arbitrary element of $\operatorname{dom} \psi_{S_1}$. We have from (16.8.2) that

$$D(S_2) - x \subset F - z,$$

from which

$$D(S_2) \subset F + x - z.$$

Using Corollary 16.7 again, we deduce from this that

$$\operatorname{dom} \psi_{S_2} \subset F + x - z,$$

i.e.,

$$x - \operatorname{dom} \psi_{S_2} \subset F + z = \overline{\operatorname{aff}(D(S_1) - D(S_2))}.$$

(16.8.1) now follows by allowing x to run through $\operatorname{dom} \psi_{S_1}$. ▌

Corollary 16.9. *Let $S\colon E \mapsto 2^{E^*}$ be maximal monotone. Then*

$$\overline{\operatorname{aff}(\operatorname{dom} \chi_S)} = \overline{\operatorname{aff}(\operatorname{dom} \psi_S)} = \overline{\operatorname{aff} D(S)}.$$

Proof. This can be proved directly from Corollary 16.7. Alternatively, it can be deduced from Theorem 16.8 by substituting $S_1 := S$ and $S_2 := N_{\{0\}}$. (We note from Theorem 16.2, for instance, that $\operatorname{dom} \chi_{S_2} = \operatorname{dom} \psi_{S_2} = D(S_2) = \{0\}$). ▌

We will also use the final result of this section in our analysis of constraint qualifications in Section 23.

Theorem 16.10. *Let F be a closed subspace of E, $T\colon E \mapsto 2^{E^*}$ be monotone and $\emptyset \neq D(T) \subset F$. Define $T_F\colon F \mapsto 2^{F^*}$ by*

$$T_F x := \{x^*|_F\colon x^* \in Tx\}.$$

Then:
(a) T_F is monotone.
(b) $D(T_F) = D(T)$.
(c) $\chi_{(T_F)} = (\chi_T)|F$ — hence $\operatorname{dom} \chi_{(T_F)} = F \cap \operatorname{dom} \chi_T$.
(d) If T is F-saturated then $\operatorname{dom} \chi_{(T_F)} = \operatorname{dom} \chi_T$.
(e) T is maximal monotone $\iff$ T is F-saturated and T_F is maximal monotone.

Proof. We leave the proofs of (a), (b) and (c) as exercises. (d) follows from (c) and Lemma 16.6 (with $w = 0$).

(e)($\Longrightarrow$) Suppose that T is maximal monotone. Since it follows from Lemma 16.5 that T is F-saturated, it only remains to prove that T_F is maximal monotone. Let $(z, z^*) \in F \times F^*$ be monotonically related to $G(T_F)$ in $F \times F^*$. It follows from the definition of T_F that

$$\text{for all } (t, t^*) \in G(T), \quad \langle t - z, t^*|_F - z^* \rangle \geq 0.$$

From the "extension form" of the Hahn–Banach theorem, Theorem 4.9, there exists $y^* \in E^*$ such that $y^*|_F = z^*$. We then have

$$\text{for all } (t,t^*) \in G(T), \quad \langle t - z, t^*|_F - y^*|_F \rangle \geq 0.$$

In the above inequality, $t \in D(T) \subset F$ and also $z \in F$. We derive that,

$$\text{for all } (t,t^*) \in G(T), \quad \langle t - z, t^* - y^* \rangle \geq 0.$$

From the maximal monotonicity of T, $(z, y^*) \in G(T)$, hence $(z, z^*) = (z, y^*|_F) \in G(T_F)$. Thus T_F is maximal monotone, as required.
(e)($\Longleftarrow$) We now assume that T is F-saturated and T_F is maximal monotone, and prove that T is maximal monotone. Suppose that $(z, z^*) \in E \times E^*$ and

$$\text{for all } (t,t^*) \in G(T), \quad \langle t - z, t^* - z^* \rangle \geq 0. \tag{16.10.1}$$

We will deduce that $(z, z^*) \in G(T)$. We first prove that

$$y^* \in F^\perp \quad \Longrightarrow \quad \langle z, y^* \rangle = 0. \tag{16.10.2}$$

Let $y^* \in F^\perp$, and fix $(t, t^*) \in G(T)$. We note, for future reference, that

$$t \in F. \tag{16.10.3}$$

Let λ be an arbitrary real number. Since T is F-saturated, $(t, t^* + \lambda y^*) \in G(T)$ thus, from (16.10.1),

$$\langle t - z, t^* + \lambda y^* - z^* \rangle \geq 0.$$

We derive from this and (16.10.3) that

$$\lambda \langle z, y^* \rangle = \lambda \langle z - t, y^* \rangle \leq \langle t - z, t^* - z^* \rangle.$$

Since this holds for all $\lambda \in \mathbb{R}$, $\langle z, y^* \rangle = 0$, which completes the proof of (16.10.2). Since F is closed, it follows from (16.10.2) and Theorem 4.4 that

$$z \in F. \tag{16.10.4}$$

Now let (t, s^*) be an arbitrary element of $G(T_F)$. (16.10.3) is again satisfied and, from the definition of T_F, there exists $t^* \in E^*$ such that $t^*|_F = s^*$ and $(t, t^*) \in G(T)$. Consequently, from (16.10.1),

$$\langle t - z, t^* - z^* \rangle \geq 0.$$

We now derive from (16.10.3) and (16.10.4) that

$$\langle t - z, t^*|_F - z^*|_F \rangle \geq 0,$$

that is to say

$$\langle t - z, s^* - z^*|_F \rangle \geq 0.$$

Since this holds for all $(t, s^*) \in G(T_F)$ and T_F is maximal monotone, $(z, z^*|_F) \in G(T_F)$. From the definition of T_F, there exists $x^* \in Tz$ such that $x^*|_F = z^*|_F$. But then

$$z^* - x^* \in F^\perp.$$

Since T is F-saturated,

$$z^* = x^* + (z^* - x^*) \in Tz + F^\perp = Tz,$$

i.e., $(z, z^*) \in G(T)$. This completes the proof that T is maximal monotone. ∎

17. A general local boundedness theorem

Let $S\colon E \mapsto 2^{E^*}$ be nontrivial and $v \in E$. We say that S is *locally bounded at* v if there exist $\theta,\ Q > 0$ such that

$$(s, s^*) \in G(S) \text{ and } \|s - v\| < \theta \quad \Longrightarrow \quad \|s^*\| \le Q.$$

Borwein–Fitzpatrick proved the following result in [10]: *Let S be nontrivial and monotone, and v be an absorbing point of $D(S)$. Then S is locally bounded at v.* This result is clearly extended by Theorem 17.1:

Theorem 17.1. *Let $S\colon E \mapsto 2^{E^*}$ be nontrivial and monotone and* co$D(S)$ *surround v. Then S is locally bounded at v.*

Theorem 17.1 follows easily from the "ψ local boundedness theorem", Theorem 17.3 below and the D–dom lemma, Lemma 15.2. The ψ local boundedness theorem does not involve monotonicity in any way. We start off with a pair of equivalences valid for any nontrivial multifunction on a Banach space. We will use Lemma 17.2(b) in the ψ local boundedness theorem, and also in Lemma 18.7, while we will use Lemma 17.2(a) as part of our proof of the six set theorem in Lemma 18.2, and also as part of our proof of Rockafellar's sum theorem in Lemma 20.3.

Lemma 17.2. *Let $S\colon E \mapsto 2^{E^*}$ be nontrivial and $v \in E$. Then:*
(a) dom χ_S *surrounds* $v \iff$ *there exist $\eta > 0$ and $n \ge 1$ such that*

$$\mu \in \mathcal{CO}(S) \quad \Longrightarrow \quad r(\mu) + n(1 + \|p(\mu)\|) \ge \langle v, q(\mu)\rangle + \eta\|q(\mu)\|. \tag{17.2.1}$$

(b) dom ψ_S *surrounds* $v \iff$ *there exist $\eta > 0$ and $n \ge 1$ such that*

$$(s, s^*) \in G(S) \quad \Longrightarrow \quad \langle s, s^*\rangle + n(1 + \|s\|) \ge \langle v, s^*\rangle + \eta\|s^*\|.$$

Proof. We give the details of (a) — (b) is similar but simpler.
($\Longrightarrow$) From the dom lemma, Lemma 12.2, there exist $\eta > 0$ and $n \ge 1$ such that

$$w \in E \text{ and } \|w\| \le \eta \quad \Longrightarrow \quad \chi_S(v + w) \le n.$$

Thus, from the definition of χ_S,

$$w \in E, \quad \|w\| \le \eta \quad \text{and} \quad \mu \in \mathcal{CO}(S) \quad \Longrightarrow$$
$$\langle v + w, q(\mu)\rangle - r(\mu) \le n(1 + \|p(\mu)\|),$$

that is to say,

$$\mu \in \mathcal{CO}(S), \quad w \in E \quad \text{and} \quad \|w\| \le \eta \quad \Longrightarrow$$
$$r(\mu) + n(1 + \|p(\mu)\|) \ge \langle v, q(\mu)\rangle + \langle w, q(\mu)\rangle.$$

We now obtain (17.2.1) by taking the supremum of the right-hand expression over all $w \in E$ such that $\|w\| \le \eta$.
($\Longleftarrow$) If η and n satisfy (17.2.1) then, reversing the above steps,

$$w \in E \text{ and } \|w\| \le \eta \quad \Longrightarrow \quad \chi_S(v + w) \le n \quad \Longrightarrow \quad v + w \in \operatorname{dom} \chi_S. \ \blacksquare$$

We now come to the ψ local boundedness theorem:

Theorem 17.3. *Let $S\colon E \mapsto 2^{E^*}$ be nontrivial, $v \in E$ and $\operatorname{dom}\psi_S$ surround v. Then S is locally bounded at v.*

Proof. Let $\eta > 0$ and $n \geq 1$ be as in Lemma 17.2(b). Then

$$\begin{aligned}(s,s^*) \in G(S) &\implies \eta\|s^*\| - \|s-v\|\|s^*\| \leq n(1+\|s\|)\\ &\implies \eta\|s^*\| - \|s-v\|\|s^*\| \leq n(1+\|s-v\|+\|v\|)\\ &\implies \eta\|s^*\| - \|s-v\|\|s^*\| \leq n(1+\|v\|)(1+\|s-v\|).\end{aligned}$$

Let $\theta := \eta/2$. Then

$$(s,s^*) \in G(S) \text{ and } \|s-v\| \leq \theta \quad\implies\quad \theta\|s^*\| \leq n(1+\|v\|)(1+\theta).$$

Thus the definition of local boundedness is satisfied with

$$Q := n(1+\|v\|)(1+\theta)/\theta. \blacksquare$$

18. The six set theorem and the nine set theorem

We start this section by recalling what is known about $D(S)$ when S is a maximal monotone multifunction on a general (i.e., possibly nonreflexive) Banach space. It is not always true that $D(S)$ is convex. In fact, we shall now give an example of a function $f \in \mathcal{PCLSC}(\mathbb{R}^2)$ such that $D(\partial f)$ is not convex. Define $f\colon \mathbb{R}^2 \mapsto \mathbb{R} \cup \{\infty\}$ by

$$f(x_1,x_2) := \begin{cases} |x_2| \vee (1-\sqrt{1-x_1{}^2}), & \text{if } |x_1| \vee |x_2| \leq 1;\\ \infty, & \text{otherwise.}\end{cases}$$

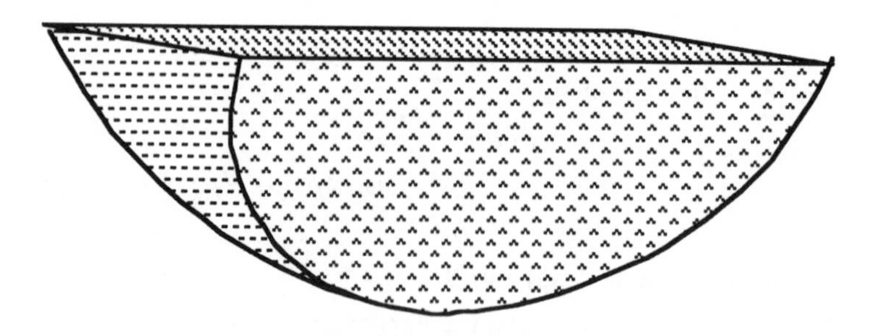

However, Rockafellar proved in [41], Theorem 1, p. 398 (see also Phelps, [35], Theorem 1.9, p. 6) that $D(S)$ is not far from being convex in the following sense: *Let S be maximal monotone and* $\operatorname{int}(\operatorname{co}D(S)) \neq \emptyset$. *Then* $\operatorname{int} D(S)$ *is convex and* $\overline{D(S)} = \overline{\operatorname{int} D(S)}$, *so* $\overline{D(S)}$ *is also convex.*

In the main result of this section, the "six set theorem", Theorem 18.3, we prove that the six sets $\operatorname{int} D(S)$, $\operatorname{int}(\operatorname{co} D(S))$, $\operatorname{int}(\operatorname{dom}\chi_S)$, $\operatorname{sur} D(S)$, $\operatorname{sur}(\operatorname{co} D(S))$ and $\operatorname{sur}(\operatorname{dom}\chi_S)$ coincide and, in its consequence, the "nine set theorem", Theorem 18.4, we prove that, if $\operatorname{sur}(\operatorname{dom}\chi_S) \neq \emptyset$, then the nine sets $\overline{D(S)}$, $\overline{\operatorname{co} D(S)}$, $\overline{\operatorname{dom}\chi_S}$, $\overline{\operatorname{int} D(S)}$, $\overline{\operatorname{int}(\operatorname{co} D(S))}$, $\overline{\operatorname{int}(\operatorname{dom}\chi_S)}$, $\overline{\operatorname{sur} D(S)}$, $\overline{\operatorname{sur}(\operatorname{co} D(S))}$ and $\overline{\operatorname{sur}(\operatorname{dom}\chi_S)}$ coincide.

The six set theorem and the nine set theorem not only extend the results referred to above, but also answer in the affirmative a question raised by Phelps (see [34], p. 29 and [35], p. 8), namely whether an absorbing point of $D(S)$ is necessarily an interior point.

As a preliminary to the six set theorem, we use the minimax technique to obtain an equivalence for general (i.e., possibly nonreflexive) Banach spaces that is similar in spirit to Lemma 10.1.

Lemma 18.1. *Let $S\colon E \mapsto 2^{E^*}$ be nontrivial. Then* (18.1.1)$\Longleftrightarrow$(18.1.2).

$$\left.\begin{array}{l}\textit{There exists } K > 0 \textit{ such that}\\ \qquad \mu \in \mathcal{CO}(S) \Longrightarrow r(\mu) + K\|p(\mu)\| \geq 0.\end{array}\right\} \tag{18.1.1}$$

$$\left.\begin{array}{l}\textit{There exists } x^* \in E^* \textit{ such that}\\ \qquad (s, s^*) \in G(S) \Longrightarrow \langle s, s^* - x^*\rangle \geq 0.\end{array}\right\} \tag{18.1.2}$$

Proof. ($\Longrightarrow$) Define $h\colon \mathbb{R}^{(E\times E^*)} \times E^* \mapsto \mathbb{R}$ by

$$h(\mu, x^*) := r(\mu) - \langle p(\mu), x^*\rangle. \tag{18.1.3}$$

Let K satisfy (18.1.1), $A := \mathcal{CO}(S)$ and $B := \{x^* \in E^*\colon \|x^*\| \leq K\}$, with the topology $w(E^*, E)$. From the Banach–Alaoglu theorem, Theorem 4.1, B is compact. If $\mu \in A$ then, from the one–dimensional Hahn–Banach theorem, Corollary 1.2,

$$\max_{x^* \in B} h(\mu, x^*) = r(\mu) + K\|p(\mu)\|,$$

hence

$$\inf_A \max_B h \geq 0.$$

The sets A and B are convex. Since h is affine in each variable and continuous on B, from the minimax theorem, Theorem 3.1,

$$\max_B \inf_A h \geq 0,$$

hence there exists $x^* \in B \subset E^*$ such that

$$\mu \in A \quad\Longrightarrow\quad r(\mu) - \langle p(\mu), x^*\rangle \geq 0.$$

(18.1.2) now follows by allowing μ to run through the values $\delta_{(s,s^*)}$ ($(s, s^*) \in G(S)$).

($\Longleftarrow$) Let x^* be as in (18.1.2). If $(s, s^*) \in G(S)$ then

$$-\langle s, s^* \rangle \leq -\langle s, x^* \rangle.$$

Let $\mu \in \mathcal{CO}(S)$. Multiplying the inequality above by $\mu(s, s^*)$ and summing up over all $(s, s^*) \in G(S)$,

$$-r(\mu) \leq -\langle p(\mu), x^* \rangle.$$

Since

$$-\langle p(\mu), x^* \rangle \leq \|p(\mu)\| \|x^*\|,$$

(18.1.1) follows with $K := \|x^*\|$. ∎

Lemma 18.2 will lead easily to the six set theorem. We note the role played in the last line of the proof by the "qualitative" Corollary 12.3.

Lemma 18.2. *Let S: $E \mapsto 2^{E^*}$ be maximal monotone. Then*

$$\operatorname{sur}(\operatorname{dom} \chi_S) \subset \operatorname{int} D(S).$$

Proof. We first prove that

$$\operatorname{dom} \chi_S \text{ surrounds } 0 \quad \Longrightarrow \quad 0 \in D(S). \tag{18.2.1}$$

If $\operatorname{dom} \chi_S$ surrounds 0 then, from Lemma 17.2(a), there exist $\eta > 0$ and $n \geq 1$ such that

$$\mu \in \mathcal{CO}(S) \quad \Longrightarrow \quad r(\mu) + n(1 + \|p(\mu)\|) \geq \eta \|q(\mu)\|. \tag{18.2.2}$$

By decreasing η if necessary, we can and will suppose that $\eta \in (0, 1]$. We now write $K := n/\eta$, and we shall prove that

$$\mu \in \mathcal{CO}(S) \quad \Longrightarrow \quad r(\mu) + K\|p(\mu)\| \geq 0. \tag{18.2.3}$$

To this end, let $\mu \in \mathcal{CO}(S)$. If $\|q(\mu)\| \leq K$ then, from the pqr–lemma, Lemma 9.1,

$$\begin{aligned} r(\mu) + K\|p(\mu)\| &\geq \langle p(\mu), q(\mu) \rangle + K\|p(\mu)\| \\ &\geq K\|p(\mu)\| - \|p(\mu)\| \|q(\mu)\| \\ &= (K - \|q(\mu)\|)\|p(\mu)\| \geq 0, \end{aligned}$$

and (18.2.3) follows. If, on the other hand, $\|q(\mu)\| > K$ then, since $K = n/\eta \geq n$,

$$r(\mu) + K\|p(\mu)\| \geq r(\mu) + n\|p(\mu)\|,$$

from (18.2.2),

$$\begin{aligned} &\geq \eta\|q(\mu)\| - n \\ &= \eta(\|q(\mu)\| - K) > 0, \end{aligned}$$

and (18.2.3) follows again. From Lemma 18.1,

$$\left.\begin{array}{l}\text{there exists } x^* \in E^* \text{ such that} \\ \quad (s, s^*) \in G(S) \Longrightarrow \langle s - 0, s^* - x^* \rangle \geq 0.\end{array}\right\} \tag{18.1.2}$$

Since S is maximal monotone,

$$(0, x^*) \in G(S).$$

This gives (18.2.1). From (18.2.1) and Lemma 15.3,

$$\operatorname{sur}(\operatorname{dom} \chi_S) \subset D(S).$$

The result follows since, from the dom corollary, Corollary 12.3, $\operatorname{sur}(\operatorname{dom} \chi_S)$ is open. ▌

Theorem 18.3. *Let S: $E \mapsto 2^{E^*}$ be maximal monotone. Then*

$$\begin{aligned} \operatorname{int} D(S) &= \operatorname{int}(\operatorname{co} D(S)) = \operatorname{int}(\operatorname{dom} \chi_S) \\ &= \operatorname{sur} D(S) = \operatorname{sur}(\operatorname{co} D(S)) = \operatorname{sur}(\operatorname{dom} \chi_S). \end{aligned}$$

Proof. From the D–dom lemma, Lemma 15.2,

$$\operatorname{int} D(S) \subset \operatorname{int}(\operatorname{co} D(S)) \subset \operatorname{int}(\operatorname{dom} \chi_S)$$

and

$$\operatorname{sur} D(S) \subset \operatorname{sur}(\operatorname{co} D(S)) \subset \operatorname{sur}(\operatorname{dom} \chi_S).$$

Obviously $\operatorname{int}(\ldots) \subset \operatorname{sur}(\ldots)$, and the result follows from Lemma 18.2. ▌

Theorem 18.4. *Let S be maximal monotone and* $\operatorname{sur}(\operatorname{dom} \chi_S) \neq \emptyset$. *Then*

$$\begin{aligned} \overline{D(S)} &= \overline{\operatorname{co} D(S)} = \overline{\operatorname{dom} \chi_S} \\ &= \overline{\operatorname{int} D(S)} = \overline{\operatorname{int}(\operatorname{co} D(S))} = \overline{\operatorname{int}(\operatorname{dom} \chi_S)} \\ &= \overline{\operatorname{sur} D(S)} = \overline{\operatorname{sur}(\operatorname{co} D(S))} = \overline{\operatorname{sur}(\operatorname{dom} \chi_S)}. \end{aligned}$$

Proof. Obviously, $\overline{\operatorname{int} D(S)} \subset \overline{D(S)}$ and, from the D–dom lemma, Lemma 15.2, $\overline{D(S)} \subset \overline{\operatorname{co} D(S)} \subset \overline{\operatorname{dom} \chi_S}$. From the dom corollary, Corollary 12.3, $\operatorname{int}(\operatorname{dom} \chi_S) = \operatorname{sur}(\operatorname{dom} \chi_S) \neq \emptyset$, hence (see, for instance, Kelly–Namioka, [28], 13.1(i), p. 100–111),

$$\overline{\operatorname{dom} \chi_S} = \overline{\operatorname{int}(\operatorname{dom} \chi_S)}.$$

Thus we have

$$\overline{\operatorname{int} D(S)} \subset \overline{D(S)} \subset \overline{\operatorname{co} D(S)} \subset \overline{\operatorname{dom} \chi_S} = \overline{\operatorname{int}(\operatorname{dom} \chi_S)}.$$

The result now follows by combining this with Theorem 18.3. ▌

Much stronger results are known if E is reflexive, and we will devote the rest of this section to discussing these, which are based ultimately on techniques introduced by Rockafellar in [42]. Theorem 18.6 extends the result proved in [42], Theorem 2, p. 89 that if E is reflexive and S: $E \mapsto 2^{E^*}$ is maximal monotone then $\overline{D(S)}$ is convex. The results of Theorems 18.6 and 18.8 can be deduced from Simons, [53], Theorem 13, p. 187 and the D–dom lemma, Lemma 15.2. Lemma 18.5 also provides a model for the proofs of Theorems 27.6 and 27.8.

Lemma 18.5. *Let E be reflexive and $S\colon E \mapsto 2^{E^*}$ be maximal monotone. Then*

$$\operatorname{dom} \psi_S \subset \overline{D(S)}.$$

Proof. We first prove that

$$0 \in \operatorname{dom} \psi_S \quad \Longrightarrow \quad 0 \in \overline{D(S)}. \tag{18.5.1}$$

So let $0 \in \operatorname{dom} \psi_S$ and $\varepsilon \in (0,1)$. We shall show that

$$\text{there exists } w \in D(S) \text{ such that } \|w\| < \varepsilon, \tag{18.5.2}$$

which will give (18.5.1). Put $M := 0 \vee \psi_S(0)$. Then $M \geq 0$ and

$$(w, w^*) \in G(S) \quad \Longrightarrow \quad -\langle w, w^* \rangle - M(1 + \|w\|) \leq 0. \tag{18.5.3}$$

Choose $\lambda > 0$ so that

$$\lambda M < \frac{\varepsilon^2}{5} < 1, \tag{18.5.4}$$

and define $T\colon E \mapsto 2^{E^*}$ by $Tx := S(\lambda x) \quad (x \in E)$. Then T is also maximal monotone hence, from Lemma 10.2(b), there exists $(x, x^*) \in G(T)$ such that

$$\|x\|^2 + \|x^*\|^2 + 2\langle x, x^* \rangle = 0. \tag{18.5.5}$$

It is evident from the proof of the perfect square trick, Lemma 7.1, that $\|x\| = \|x^*\|$. Thus, from (18.5.5), $\langle x, x^* \rangle = -\|x\|^2$. Setting $w := \lambda x$, we derive that

$$-\langle w, x^* \rangle = -\lambda \langle x, x^* \rangle = \lambda \|x\|^2 = \|w\|^2/\lambda \geq 0. \tag{18.5.6}$$

Since $(x, x^*) \in G(T)$, $(w, x^*) \in G(S)$ and so, substituting $w^* := x^*$ in (18.5.3),

$$\frac{\|w\|^2}{\lambda} - M(1 + \|w\|) \leq 0,$$

i.e.,

$$\|w\|^2 - \lambda M \|w\| - \lambda M \leq 0$$

thus, completing the square and using (18.5.4),

$$\begin{aligned}
\|w\| &\leq \frac{\lambda M + \sqrt{\lambda^2 M^2 + 4\lambda M}}{2} \\
&\leq \sqrt{\lambda^2 M^2 + 4\lambda M} \\
&= \sqrt{\lambda M(\lambda M + 4)} \\
&< \sqrt{\frac{\varepsilon^2}{5}(1 + 4)} \\
&= \varepsilon.
\end{aligned}$$

Since $w \in D(S)$, this establishes (18.5.2), and consequently (18.5.1). The result now follows from Lemma 15.3. ▌

Theorem 18.6. *Let E be reflexive and $S: E \mapsto 2^{E^*}$ be maximal monotone. Then*

$$\overline{D(S)} = \overline{\mathrm{co}D(S)} = \overline{\mathrm{dom}\,\chi_S} = \overline{\mathrm{dom}\,\psi_S}.$$

Proof. This is immediate from Lemma 18.5 and the D–dom Lemma, Lemma 15.2. ▌

Lemma 18.7. *Let E be reflexive and $S: E \mapsto 2^{E^*}$ be maximal monotone. Then*

$$\mathrm{int}\,(\mathrm{dom}\,\psi_S) \subset \mathrm{int}\,D(S).$$

Proof. We first prove that

$$0 \in \mathrm{int}\,(\mathrm{dom}\,\psi_S) \quad\Longrightarrow\quad 0 \in D(S). \tag{18.7.1}$$

If $0 \in \mathrm{int}\,(\mathrm{dom}\,\psi_S)$ then, from Lemma 17.2(b), there exist $\eta > 0$ and $n \geq 1$ such that

$$(s, s^*) \in G(S) \quad\Longrightarrow\quad \langle s, s^* \rangle + n(1 + \|s\|) \geq \eta\|s^*\|. \tag{18.7.2}$$

We now write $K := 2n/\eta$, and we shall prove that

$$\mu \in \mathcal{CO}(S) \quad\Longrightarrow\quad r(\mu) + K\|p(\mu)\| \geq 0. \tag{18.7.3}$$

To this end, let $\mu \in \mathcal{CO}(S)$ and $\varepsilon \in (0, 1)$. Since $0 \in \mathrm{dom}\,\psi_S$, it follows from the proofs of (18.5.2) and (18.5.6) that there exists $(w, x^*) \in G(S)$ such that $\|w\| < \varepsilon$ and $\langle w, x^* \rangle \leq 0$. Putting $(s, s^*) := (w, x^*)$ in (18.7.2),

$$\eta\|x^*\| \leq \langle w, x^* \rangle + n(1 + \|w\|) \leq 2n,$$

and so

$$\|x^*\| \leq K.$$

Since S is monotone and $(w, x^*) \in G(S)$,

$$r(\mu) - \langle p(\mu), x^* \rangle - \langle w, q(\mu) \rangle + \langle w, x^* \rangle \geq 0$$

hence

$$r(\mu) + K\|p(\mu)\| + \varepsilon\|q(\mu)\| + \varepsilon K \geq 0.$$

We now obtain (18.7.3) by letting $\varepsilon \to 0$. The rest of the proof follows exactly the same lines as that of Lemma 18.2 (starting from (18.2.3)). ▌

Theorem 18.8. *Let E be reflexive and $S: E \mapsto 2^{E^*}$ be maximal monotone. Then*

$$\mathrm{int}\,D(S) = \mathrm{int}\,(\mathrm{co}D(S)) = \mathrm{int}\,(\mathrm{dom}\,\chi_S) = \mathrm{int}\,(\mathrm{dom}\,\psi_S).$$

Proof. This is immediate from Lemma 18.7 and the D–dom Lemma, Lemma 15.2. ▌

Theorems 18.4 and 18.6 suggest the following problem:

Problem 18.9. Is $\overline{D(S)}$ necessarily convex when E is not reflexive, S is maximal monotone and $\operatorname{sur}(\operatorname{dom}\chi_S) = \emptyset$? (See Section 26 for more on this problem.)

Theorem 18.3 and Lemma 18.7 suggest the following problem:

Problem 18.10. Let $S\colon E \mapsto 2^{E^*}$ be maximal monotone. If E is not reflexive, do we always have:

$$\operatorname{int}(\operatorname{dom}\psi_S) \subset D(S)?$$

19. The range of a sum

Let $S_1\colon E \mapsto 2^{E^*}$ and $S_2\colon E \mapsto 2^{E^*}$ be monotone. *Brézis–Haraux appproximation* is concerned with finding conditions under which $R(S_1 + S_2) \approx R(S_1) + R(S_2)$, which means that

$$\left.\begin{array}{c} \overline{R(S_1 + S_2)} = \overline{R(S_1) + R(S_2)} \\ \text{and} \\ \operatorname{int}\big[R(S_1 + S_2)\big] = \operatorname{int}\big[R(S_1) + R(S_2)\big]. \end{array}\right\} \tag{19.0.1}$$

Brézis and Haraux introduced this concept in Hilbert spaces in [13], and gave applications to Hammerstein integral equations, partial differential equations with nonlinear boundary conditions, and nonlinear periodic equations of evolution. They prove the following result in [13], Théorème 3, p. 173: *Let E be a Hilbert space, $S_1\colon E \mapsto 2^{E^*}$ and $S_2\colon E \mapsto 2^{E^*}$ be monotone, and $S_1 + S_2$ be maximal monotone. If either,*

$$\left.\begin{array}{l} \text{for all } (x_1, x_1^*) \in D(S_1) \times R(S_1) \text{ and } (x_2, x_2^*) \in D(S_2) \times R(S_2), \\ \qquad \sup_{(w, w_1^*) \in G(S_1)} \langle w - x_1, x_1^* - w_1^* \rangle < \infty \\ \qquad \text{and} \\ \qquad\qquad \sup_{(w, w_2^*) \in G(S_2)} \langle w - x_2, x_2^* - w_2^* \rangle < \infty. \end{array}\right\} \tag{19.0.2}$$

or

$$\left.\begin{array}{l} D(S_1) \subset D(S_2) \text{ and, for all } (x_2, x_2^*) \in D(S_2) \times R(S_2), \\ \qquad \sup_{(w, w_2^*) \in G(S_2)} \langle w - x_2, x_2^* - w_2^* \rangle < \infty. \end{array}\right\} \tag{19.0.3}$$

Then (19.0.1) *is satisfied.* These results were extended by Reich in [38], Theorem 2.2, p. 315 to the case where E is a reflexive Banach spaces in which J and J^{-1} are single–valued.

We now introduce the function ξ_S "dual" to the function ψ_S defined in Definition 15.1. We will show in Lemma 19.3 and Theorem 19.4 how the use of ξ_S leads to generalizations of the results described above. We do not make the assumption that J or J^{-1} is single–valued. We will return to our consideration of the function ξ_S in Section 27.

Definition 19.1. If $S\colon E \mapsto 2^{E^*}$ is nontrivial and monotone, we define $\xi_S\colon E^* \mapsto \mathbb{R} \cup \{\infty\}$ by

$$\xi_S(x^*) := \sup_{(w,w^*) \in G(S)} \frac{\langle w, x^* - w^* \rangle}{1 + \|w^*\|}.$$

ξ_S is convex and $w(E^*, E)$–lower semicontinuous. Using an argument analogous to that in the D–dom lemma, Lemma 15.2,

$$R(S) \subset \operatorname{dom} \xi_S. \tag{19.1.1}$$

If E is reflexive and $S\colon E \mapsto 2^{E^*}$ is maximal monotone and we define $T\colon E^* \mapsto 2^{E^{**}}$ by $Tx^* := \widehat{S^{-1}x^*}$ then $\xi_S = \psi_T$, and so Theorem 19.2 follows from Theorems 18.6 and 18.8.

Theorem 19.2. *Let E be reflexive and $S\colon E \mapsto 2^{E^*}$ be maximal monotone. Then*

$$\overline{R(S)} = \overline{\operatorname{dom} \xi_S} \quad \text{and} \quad \operatorname{int} R(S) = \operatorname{int}(\operatorname{dom} \xi_S).$$

The connection between ξ_S and Brézis-Haraux approximation is as follows:

Lemma 19.3. *Let E be reflexive, $S_1\colon E \mapsto 2^{E^*}$ and $S_2\colon E \mapsto 2^{E^*}$ be monotone, $S_1 + S_2$ be maximal monotone, and suppose that*

$$R(S_1) + R(S_2) \subset \operatorname{dom} \xi_{S_1+S_2}. \tag{19.3.1}$$

Then (19.0.1) is satisfied.

Proof. It follows from Theorem 19.2 that

$$\overline{R(S_1) + R(S_2)} \subset \overline{\operatorname{dom} \xi_{S_1+S_2}} = \overline{R(S_1 + S_2)}$$

and

$$\operatorname{int}\big[R(S_1) + R(S_2)\big] \subset \operatorname{int}(\operatorname{dom} \xi_{S_1+S_2}) = \operatorname{int}\big[R(S_1 + S_2)\big].$$

The reverse inclusions in (19.0.1) are clear since

$$R(S_1 + S_2) \subset R(S_1) + R(S_2). \blacksquare$$

Lemma 19.3 shows that Theorem 19.4 below, which requires neither the reflexivity of E nor the maximal monotonicity of $S_1 + S_2$, leads to the results from [13] and [38] referred to above.

Theorem 19.4. *Let E be a non–trivial Banach space, $S_1\colon E \mapsto 2^{E^*}$ and $S_2\colon E \mapsto 2^{E^*}$ be monotone, and $D(S_1) \cap D(S_2) \neq \emptyset$. Then (19.0.2)$\Longrightarrow$(19.3.1) and (19.0.3)$\Longrightarrow$(19.3.1).*

Proof. We introduce the intermediate condition

$$\left.\begin{aligned}&\text{for all } x_1^* \in R(S_1) \text{ and } x_2^* \in R(S_2),\\ &\qquad\text{there exists } z \in E \text{ such that}\\ &\qquad\qquad \sup_{(w,w_1^*)\in G(S_1)} \langle w - z, x_1^* - w_1^* \rangle < \infty \quad \text{and}\\ &\qquad\qquad\qquad \sup_{(w,w_2^*)\in G(S_2)} \langle w - z, x_2^* - w_2^* \rangle < \infty,\end{aligned}\right\} \tag{19.4.1}$$

and we shall obtain the required result by showing that (19.0.2)$\Longrightarrow$(19.4.1), (19.0.3)$\Longrightarrow$(19.4.1) and (19.4.1)$\Longrightarrow$(19.3.1).

((19.0.2)$\Longrightarrow$(19.4.1)) Let $z \in D(S_1) \cap D(S_2)$. (19.4.1) follows from (19.0.2) with $x_1 = x_2 = z$. (This value of z works for any $x_1^* \in R(S_1)$ and $x_2^* \in R(S_2)$.)

((19.0.3)$\Longrightarrow$(19.4.1)) Suppose that $x_1^* \in R(S_1)$ and $x_2^* \in R(S_2)$. Choose $z \in E$ so that $(z, x_1^*) \in G(S_1)$. Then $z \in D(S_1) \subset D(S_2)$. (19.4.1) follows from (19.0.3) since, using the monotonicity of S_1,

$$\sup_{(w,w_1^*)\in G(S_1)} \langle w - z, x_1^* - w_1^* \rangle \le 0.$$

((19.4.1)$\Longrightarrow$(19.3.1)) Suppose that $x_1^* \in R(S_1)$ and $x_2^* \in R(S_2)$. Choose z as in (19.4.1), and let

$$M := \sup_{(w,w_1^*)\in G(S_1)} \langle w - z, x_1^* - w_1^* \rangle + \sup_{(w,w_2^*)\in G(S_2)} \langle w - z, x_2^* - w_2^* \rangle$$

and

$$\beta := \big[M + \langle z, x_1^* + x_2^* \rangle\big] \vee \|z\|.$$

Now let (w, w^*) be an arbitrary element of $G(S_1 + S_2)$. Then there exist $w_1^* \in S_1 w$ and $w_2^* \in S_2 w$ such that $w_1^* + w_2^* = w^*$. We have

$$\begin{aligned}\langle w, x_1^* + x_2^* - w^* \rangle &= \langle w - z, x_1^* + x_2^* - w^* \rangle + \langle z, x_1^* + x_2^* - w^* \rangle\\ &= \langle w - z, x_1^* - w_1^* \rangle + \langle w - z, x_2^* - w_2^* \rangle\\ &\qquad\qquad + \langle z, x_1^* + x_2^* - w^* \rangle.\end{aligned}$$

Thus, by hypothesis,

$$\begin{aligned}\langle w, x_1^* + x_2^* - w^* \rangle &\le M + \langle z, x_1^* + x_2^* - w^* \rangle\\ &\le M + \langle z, x_1^* + x_2^* \rangle + \|z\|\|w^*\|\\ &\le \beta(1 + \|w^*\|).\end{aligned}$$

Dividing by $(1 + \|w^*\|)$ and taking the supremum over $(w, w^*) \in G(S_1 + S_2)$, we obtain $\xi_{S_1+S_2}(x_1^* + x_2^*) \le \beta$, from which $x_1^* + x_2^* \in \operatorname{dom} \xi_{S_1+S_2}$. Since this holds for any $x_1^* \in R(S_1)$ and $x_2^* \in R(S_2)$, we have established (19.3.1). ∎

Remark 19.5. There is a another condition which implies (19.0.1). This condition, due to Pazy, appears in the appendix of [13] for Hilbert spaces. Pazy's results were generalized (with applications) by Reich in [38], Proposition 2.3, p. 315–316 to the case where E is a reflexive Banach spaces in which J and J^{-1} are single–valued. By analogy with Theorem 19.4, one might hope that (19.3.1) is also true in the situation considered there. However, we do not know if this is the case. After some simple transformations, we can reduce this question to the following problem, which we have stated in a Hilbert space for simplicity.

Problem 19.6. Let H be a Hilbert space and $S\colon H \mapsto 2^H$ and $T\colon H \mapsto 2^H$ be monotone. Suppose that $R(S) + R(T) \ni 0$, and also that there exists $M \geq 0$ such that, for arbitrarily small $\lambda > 0$, there exists $(u_\lambda, u_\lambda^*) \in G(S)$ such that $\|u_\lambda^*\| \leq M$ and $(u_\lambda, -u_\lambda^* - \lambda u_\lambda) \in G(T)$. Then does there necessarily exist $\beta \geq 0$ such that

$$(w, w^*) \in G(S+T) \quad \Longrightarrow \quad \langle w, w^* \rangle + \beta(1 + \|w^*\|) \geq 0?$$

V. The sum problem for reflexive spaces

20. The maximal monotonicity of a sum

If $S_1\colon E \mapsto 2^{E^*}$ and $S_2\colon E \mapsto 2^{E^*}$ are nontrivial and monotone and $D(S_1) \cap D(S_2) \neq \emptyset$ then $S_1 + S_2$, defined in (8.0.1), is obviously nontrivial and monotone. On the other hand, it does *not* follow that if S_1 and S_2 are maximal monotone and $D(S_1) \cap D(S_2) \neq \emptyset$ then $S_1 + S_2$ is maximal monotone. As an example of this, let C_1 and C_2 be two closed disks in the plane that touch at the point p as in the diagram below.

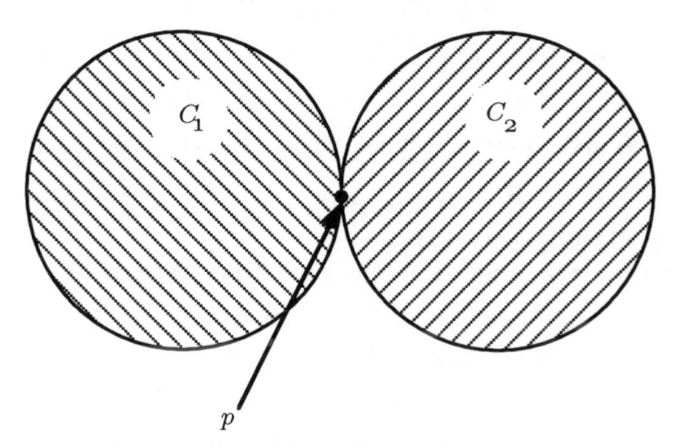

Then N_{C_1} and N_{C_2} (see (8.1.1)) are maximal monotone. Note that

$$(N_{C_1} + N_{C_2})(0) = N_{C_1}(0) + N_{C_2}(0),$$

which can be represented by the diagram

which is a proper subset of $\mathbb{R}^2$. Thus $G(N_{C_1} + N_{C_2})$ is a proper subset of

$$G(N_{\{p\}}) = \{p\} \times \mathbb{R}^2,$$

and so $N_{C_1} + N_{C_2}$ is not maximal monotone. (See Phelps, [34], p. 54.) Determining conditions on S_1 and S_2 (normally called "constraint qualifications") that ensure that S_1+S_2 is maximal monotone is one of the fundamental questions in the theory of monotone multifunctions. Very little is known about this if E is a general Banach space. Much more is known if E is reflexive, so for the rest of this chapter we shall suppose that this is the case. The original milestone result due to Rockafellar (see [43], Theorem 1, p. 76) was that $S_1 + S_2$ is maximal monotone if

$$D(S_1) \cap \operatorname{int} D(S_2) \neq \emptyset. \tag{20.0.1}$$

Taking into account the D-dom lemma, Lemma 15.2, in order to establish this, it is enough to prove that $S_1 + S_2$ is maximal monotone if

$$\operatorname{dom} \chi_{S_1} \cap \operatorname{int} \operatorname{dom} \chi_{S_2} \neq \emptyset.$$

This is, indeed, the result of Theorem 20.5. Apart from this, we will also set up in this section the machinery that we shall need for more general results that we will present later on in the chapter.

In order to gain insight into the processes involved in this problem, let us work backwards and suppose for the moment that $S_1 + S_2$ is maximal monotone. From Lemma 10.2(b), there exists $(z, z^*) \in G(S_1 + S_2)$ such that

$$\|z\|^2 + \|z^*\|^2 + 2\langle z, z^*\rangle = 0.$$

Let $(z, x_1^*) \in G(S_1)$, $(z, x_2^*) \in G(S_2)$ and $x_1^* + x_2^* = z^*$. So we have

$$\|z\|^2 + \|x_1^* + x_2^*\|^2 + 2\langle z, x_1^* + x_2^*\rangle = 0$$

and, since S_1 and S_2 are monotone, for all $(s_1, s_1^*) \in G(S_1)$ and $(s_2, s_2^*) \in G(S_2)$,

$$\langle s_1 - z, s_1^* - x_1^*\rangle \geq 0 \quad \text{and} \quad \langle s_2 - z, s_2^* - x_2^*\rangle \geq 0.$$

Combining all this together, we see that if S_1, S_2 and $S_1 + S_2$ are maximal monotone then there exist $z \in E$, and x_1^* and $x_2^* \in E^*$ such that

$$\begin{aligned} (s_1, s_1^*) \in G(S_1) \text{ and } (s_2, s_2^*) \in G(S_2) \quad &\Longrightarrow \\ 2\langle s_1 - z, s_1^* - x_1^*\rangle + 2\langle s_2 - z, s_2^* - x_2^*\rangle& \\ \geq \|z\|^2 + \|x_1^* + x_2^*\|^2 + 2\langle z, x_1^* + x_2^*\rangle&. \end{aligned}$$

The pivotal result in our investigation of the sum problem is the γ–lemma, Lemma 20.1, in which we use the minimax technique in the form of the fg–theorem, Theorem 7.2, to transform the above problem of finding one element of E and two elements of E^* into the problem of finding the one scalar constant γ. Though the important point in the γ–lemma is that (20.1.1) implies (20.1.2), the fact that it is an *equivalence* and the above discussion show that these conditions are not "pulled out of a hat". The problem is, of course, to find a value of γ satisfying (20.1.1). This is achieved in Lemma 20.3 by using the dom lemma (via Lemma 17.2(a)), and in Lemma 21.1 by using the dom–dom lemma, both of which depends ultimately on Baire's theorem. In Lemma 24.2(a), we shall discuss another way in which (20.1.1) can be established, this time without the use of Baire's theorem. For future reference, we shall call (20.1.1) the "γ–condition".

Lemma 20.1. *Let E be reflexive and $S_1\colon E \mapsto 2^{E^*}$ and $S_2\colon E \mapsto 2^{E^*}$ be nontrivial. Then the conditions (20.1.1) and (20.1.2) are equivalent.*

$$\left.\begin{aligned}&\textit{There exists } \gamma \geq 0 \textit{ such that}\\&\quad(\mu_1,\mu_2) \in \mathcal{CO}(S_1)\times\mathcal{CO}(S_2) \quad\Longrightarrow\\&\qquad 2r(\mu_1+\mu_2) + 2\gamma\|p(\mu_2-\mu_1)\|\\&\qquad\qquad + \|p(\mu_1)\|^2 + \|q(\mu_1+\mu_2)\|^2 \geq 0.\end{aligned}\right\}\tag{20.1.1}$$

$$\left.\begin{aligned}&\textit{There exist } x_1^*,\ x_2^* \in E^* \textit{ and } z \in E \textit{ such that}\\&\quad(s_1,s_1^*) \in G(S_1) \textit{ and } (s_2,s_2^*) \in G(S_2) \quad\Longrightarrow\\&\qquad 2\langle s_1 - z, s_1^* - x_1^*\rangle + 2\langle s_2 - z, s_2^* - x_2^*\rangle\\&\qquad\qquad \geq \|z\|^2 + \|x_1^* + x_2^*\|^2 + 2\langle z, x_1^* + x_2^*\rangle.\end{aligned}\right\}\tag{20.1.2}$$

Proof. We shall establish the equivalence of (20.1.1) and (20.1.2) by proving their equivalence with the intermediate conditions (20.1.3) — (20.1.6) below:

$$\left.\begin{aligned}&\text{There exist } (z,z^*) \in E\times E^* \text{ and } \gamma \geq 0 \text{ such that}\\&\quad(\mu_1,\mu_2) \in \mathcal{CO}(S_1)\times\mathcal{CO}(S_2) \quad\Longrightarrow\\&\qquad 2r(\mu_1+\mu_2) + 2\gamma\|p(\mu_2-\mu_1)\| - 2\langle(p(\mu_1),z^*\rangle\\&\qquad\qquad - 2\langle z, q(\mu_1+\mu_2)\rangle - \|z\|^2 - \|z^*\|^2 \geq 0.\end{aligned}\right\}\tag{20.1.3}$$

$$\left.\begin{aligned}&\text{There exist } (z,z^*) \in E\times E^* \text{ and } x_2^* \in E^* \text{ such that}\\&\quad(\mu_1,\mu_2) \in \mathcal{CO}(S_1)\times\mathcal{CO}(S_2) \quad\Longrightarrow\\&\qquad 2r(\mu_1+\mu_2) - 2\langle p(\mu_2-\mu_1), x_2^*\rangle - 2\langle p(\mu_1), z^*\rangle\\&\qquad\qquad - 2\langle z, q(\mu_1+\mu_2)\rangle - \|z\|^2 - \|z^*\|^2 \geq 0.\end{aligned}\right\}\tag{20.1.4}$$

$$\left.\begin{aligned}&\text{There exist } x_1^*,\ x_2^* \in E^* \text{ and } z \in E \text{ such that}\\&\quad(\mu_1,\mu_2) \in \mathcal{CO}(S_1)\times\mathcal{CO}(S_2) \quad\Longrightarrow\\&\qquad 2r(\mu_1+\mu_2) - 2\langle p(\mu_1), x_1^*\rangle - 2\langle p(\mu_2), x_2^*\rangle\\&\qquad\qquad - 2\langle z, q(\mu_1+\mu_2)\rangle - \|z\|^2 - \|x_1^*+x_2^*\|^2 \geq 0.\end{aligned}\right\}\tag{20.1.5}$$

$$\left.\begin{aligned}&\text{There exist } x_1^*,\ x_2^* \in E^* \text{ and } z \in E \text{ such that}\\&\quad(s_1,s_1^*) \in G(S_1) \text{ and } (s_2,s_2^*) \in G(S_2) \quad\Longrightarrow\\&\qquad 2\langle s_1, s_1^*\rangle + 2\langle s_2, s_2^*\rangle - 2\langle s_1, x_1^*\rangle - 2\langle s_2, x_2^*\rangle\\&\qquad\qquad - 2\langle z, s_1^*+s_2^*\rangle - \|z\|^2 - \|x_1^*+x_2^*\|^2 \geq 0.\end{aligned}\right\}\tag{20.1.6}$$

((20.1.1)$\Longleftrightarrow$(20.1.3)) We write $F := E \times E^*$ with

$$\|(x,x^*)\| := \sqrt{\|x\|^2 + \|x^*\|^2}$$

and, for all $(\mu_1,\mu_2) \in \mathcal{CO}(S_1)\times\mathcal{CO}(S_2)$,

$$f(\mu_1,\mu_2) := 2r(\mu_1+\mu_2) + 2\gamma\|p(\mu_2-\mu_1)\|$$

and

$$g(\mu_1, \mu_2) := (p(\mu_1), q(\mu_1 + \mu_2)).$$

Then (20.1.1) reduces to (7.2.1), with $A := \mathcal{CO}(S_1) \times \mathcal{CO}(S_2)$. It follows from the fg–theorem, Theorem 7.2, that (20.1.1) is equivalent to:

$$\left.\begin{array}{l} \text{There exist } \gamma \geq 0 \text{ and } y^* \in F^* \text{ such that} \\ \quad (\mu_1, \mu_2) \in \mathcal{CO}(S_1) \times \mathcal{CO}(S_2) \quad \Longrightarrow \\ \qquad 2r(\mu_1 + \mu_2) + 2\gamma\|p(\mu_2 - \mu_1)\| \\ \qquad\quad - 2\big\langle (p(\mu_1), q(\mu_1 + \mu_2)), y^* \big\rangle - \|y^*\|^2 \geq 0, \end{array}\right\}$$

which is equivalent to (20.1.3)) since any element y^* of F^* can be written in the form $(z^*, \widehat{z})$ for some $(z, z^*) \in E \times E^*$, and $\|y^*\| = \sqrt{\|z\|^2 + \|z^*\|^2}$.
((20.1.3)$\Longleftrightarrow$(20.1.4)) For fixed $(z, z^*) \in E \times E^*$, we define

$$h\colon \big(\mathcal{CO}(S_1) \times \mathcal{CO}(S_2)\big) \times E^* \mapsto \mathbb{R}$$

by

$$\begin{aligned} h((\mu_1, \mu_2), x^*) :=& \\ & 2r(\mu_1 + \mu_2) - 2\langle p(\mu_2 - \mu_1), x^* \rangle - 2\langle p(\mu_1), z^* \rangle \\ & \quad - 2\langle z, q(\mu_1 + \mu_2) \rangle - \|z\|^2 - \|z^*\|^2. \end{aligned}$$

Then the implication in (20.1.3) is simply that

$$\inf_A \max_B h \geq 0,$$

where $A := \mathcal{CO}(S_1) \times \mathcal{CO}(S_2)$, and $B := \{x^* \in E^*\colon \|x^*\| \leq \gamma\}$ with the topology $w(E^*, E)$. From the Banach–Alaoglu theorem, Theorem 4.1, B is compact. The sets A and B are convex, h is affine on A, and affine and continuous on B. Thus, from the minimax theorem, Theorem 3.1, this is, in turn, equivalent to

$$\max_B \inf_A h \geq 0.$$

This implies that (20.1.3) is equivalent to (20.1.4).

((20.1.4)$\Longleftrightarrow$(20.1.5)) This follows from the substitution $x_1^* := z^* - x_2^*$.

((20.1.5)$\Longleftrightarrow$(20.1.6)) If x_1^*, x_2^* and z satisfy (20.1.5) then (20.1.6) follows by restricting μ_1 and μ_2 to the values $\delta_{(s_1, s_1^*)}$ and $\delta_{(s_2, s_2^*)}$. If, conversely, (20.1.6) is satisfied and $(\mu_1, \mu_2) \in \mathcal{CO}(S_1) \times \mathcal{CO}(S_2)$ then (20.1.5) follows by multiplying the left side of the inequality in (20.1.6) by $\mu_1(s_1, s_1^*)\mu_2(s_2, s_2^*)$ and summing up over all $(s_1, s_1^*) \in G(S_1)$ and $(s_2, s_2^*) \in G(S_2)$.

((20.1.6)$\Longleftrightarrow$(20.1.2)) This can be seen by rearranging the terms and adding $\pm 2\langle z, x_1^* + x_2^* \rangle$ to both sides.

This completes the proof of Lemma 20.1. ∎

The technique used to prove Lemma 20.2 below is derived from a technique due to Minty and Browder (see [15], Lemma 6, p. 99).

Lemma 20.2. *Let E be reflexive and $S_1\colon E \mapsto 2^{E^*}$ and $S_2\colon E \mapsto 2^{E^*}$ be maximal monotone and satisfy the γ–condition, (20.1.1). Then there exists $(z, z^*) \in G(S_1 + S_2)$ such that*

$$\|z\|^2 + \|z^*\|^2 + 2\langle z, z^*\rangle = 0. \tag{20.2.1}$$

Proof. It follows from the γ–lemma, Lemma 20.1, that there exist $x_1^*,\ x_2^* \in E^*$ and $z \in E$ so that, for all $(s_1, s_1^*) \in G(S_1)$ and $(s_2, s_2^*) \in G(S_2)$,

$$2\langle s_1 - z, s_1^* - x_1^*\rangle + 2\langle s_2 - z, s_2^* - x_2^*\rangle \geq \|z\|^2 + \|x_1^* + x_2^*\|^2 + 2\langle z, x_1^* + x_2^*\rangle.$$

Putting $z^* := x_1^* + x_2^*$, this can be rewritten

$$2\langle s_1 - z, s_1^* - x_1^*\rangle + 2\langle s_2 - z, s_2^* - x_2^*\rangle \geq \|z\|^2 + \|z^*\|^2 + 2\langle z, z^*\rangle.$$

Taking the infimum over all $(s_1, s_1^*) \in G(S_1)$ and $(s_2, s_2^*) \in G(S_2)$ and using the perfect square trick, Lemma 7.1, with $x := z$ and $x^* := z^*$,

$$\begin{aligned} 2\inf_{(s,s^*)\in G(S_1)}\langle s - z, s^* - x_1^*\rangle + 2\inf_{(s,s^*)\in G(S_2)}\langle s - z, s^* - x_2^*\rangle \\ \geq \|z\|^2 + \|z^*\|^2 + 2\langle z, z^*\rangle \\ \geq 0. \end{aligned}$$

From Lemma 8.1(c), both these infima are zero hence, from Lemma 8.1(b), $(z, x_1^*) \in G(S_1)$ and $(z, x_2^*) \in G(S_2)$. Consequently,

$$(z, z^*) = (z, x_1^* + x_2^*) \in G(S_1 + S_2).$$

This completes the proof of Lemma 20.2. ∎

In Lemma 20.3, which assumes neither the reflexivity of E nor the maximality of S_1 or S_2, we will give a sufficient condition for the γ–condition, (20.1.1), to be satisfied.

Lemma 20.3. *Let $S_1\colon E \mapsto 2^{E^*}$ and $S_2\colon E \mapsto 2^{E^*}$ be nontrivial and monotone, and $\operatorname{dom}\chi_{S_1} \cap \operatorname{int}\operatorname{dom}\chi_{S_2} \neq \emptyset$. Then:*

$$\left.\begin{aligned} &\text{there exists } \gamma \geq 0 \text{ such that} \\ &\quad (\mu_1, \mu_2) \in \mathcal{CO}(S_1) \times \mathcal{CO}(S_2) \implies \\ &\quad\quad 2r(\mu_1 + \mu_2) + 2\gamma\|p(\mu_2 - \mu_1)\| \\ &\quad\quad\quad + \|p(\mu_1)\|^2 + \|q(\mu_1 + \mu_2)\|^2 \geq 0. \end{aligned}\right\} \tag{20.1.1}$$

Proof. To simplify the expressions in what follows, we write χ_i instead of χ_{S_i}. Let $v \in \operatorname{dom}\chi_1 \cap \operatorname{int}\operatorname{dom}\chi_2$. From Lemma 17.2(a), there exist $n \geq 1$ and $\eta \in (0, 1]$ such that

$$\left.\begin{aligned} &\mu_2 \in \mathcal{CO}(S_2) \implies \\ &\quad r(\mu_2) + n(1 + \|p(\mu_2)\|) \geq \langle v, q(\mu_2)\rangle + \eta\|q(\mu_2)\|. \end{aligned}\right\} \tag{20.3.1}$$

By increasing n if necessary, we can also suppose that

$$n \geq \chi_1(v) \quad \text{and} \quad n \geq \|v\|.$$

We write $\gamma := 5n^2/\eta$, and we shall show that γ has the required property. So let $(\mu_1, \mu_2) \in \mathcal{CO}(S_1) \times \mathcal{CO}(S_2)$. If $\|q(\mu_2)\| \leq \gamma$, then the inequality in (20.1.1) follows from Corollary 9.2(b), so we only have to consider the other alternative, namely that $\|q(\mu_2)\| > \gamma$. We derive then from (20.3.1) that

$$r(\mu_2) + n(1 + \|p(\mu_2)\|) - \langle v, q(\mu_2)\rangle - 5n^2 \geq 0. \tag{20.3.2}$$

Since $\chi_1(v) \leq n$,

$$r(\mu_1) + n(1 + \|p(\mu_1)\|) - \langle v, q(\mu_1)\rangle \geq 0. \tag{20.3.3}$$

Adding (20.3.2) and (20.3.3) and using the fact that

$$\|p(\mu_2)\| \leq \|p(\mu_1)\| + \|p(\mu_2 - \mu_1)\|,$$

we obtain

$$r(\mu_1 + \mu_2) + n\|p(\mu_2 - \mu_1)\| + 2n\|p(\mu_1)\| - \langle v, q(\mu_1 + \mu_2)\rangle - 3n^2 \geq 0.$$

Multiplying by 2 and using the fact that $\|v\| \leq n$,

$$2r(\mu_1 + \mu_2) + 2n\|p(\mu_2 - \mu_1)\| + 4n\|p(\mu_1)\| + 2n\|q(\mu_1 + \mu_2)\| - 6n^2 \geq 0.$$

Since $4n\|p(\mu_1)\| \leq 4n^2 + \|p(\mu_1)\|^2$ and $2n\|q(\mu_1 + \mu_2)\| \leq n^2 + \|q(\mu_1 + \mu_2)\|^2$,

$$2r(\mu_1 + \mu_2) + 2n\|p(\mu_2 - \mu_1)\| + \|p(\mu_1)\|^2 + \|q(\mu_1 + \mu_2)\|^2 - n^2 \geq 0.$$

The inequality in (20.1.1) follows from this since $n \leq \gamma$. ∎

In the next lemma, we combine the results of the two previous lemmas, and also carry out some simple bootstrapping steps.

Lemma 20.4. *Let E be reflexive.*
(a) Let $S_1\colon E \mapsto 2^{E^*}$ and $S_2\colon E \mapsto 2^{E^*}$ be maximal monotone and

$$\operatorname{dom} \chi_{S_1} \cap \operatorname{int} \operatorname{dom} \chi_{S_2} \neq \emptyset. \tag{20.4.1}$$

Then there exists $(z, z^*) \in G(S_1 + S_2)$ such that

$$\|z\|^2 + \|z^*\|^2 + 2\langle z, z^*\rangle = 0.$$

(b) Let $T_1\colon E \mapsto 2^{E^*}$ and $T_2\colon E \mapsto 2^{E^*}$ be maximal monotone and

$$\operatorname{dom} \chi_{T_1} \cap \operatorname{int} \operatorname{dom} \chi_{T_2} \neq \emptyset. \tag{20.4.2}$$

Suppose also that $w \in E$. Then there exists $(\zeta, z^*) \in G(T_1 + T_2)$ such that

$$\|\zeta - w\|^2 + \|z^*\|^2 + 2\langle \zeta - w, z^*\rangle = 0.$$

(c) Let $S_1\colon E \mapsto 2^{E^*}$ and $S_2\colon E \mapsto 2^{E^*}$ be maximal monotone and

$$\operatorname{dom} \chi_{S_1} \cap \operatorname{int} \operatorname{dom} \chi_{S_2} \neq \emptyset.$$

Suppose also that $w \in E$ and $w^* \in E^*$. Then there exists $(\zeta, \zeta^*) \in G(S_1 + S_2)$ such that

$$\|\zeta - w\|^2 + \|\zeta^* - w^*\|^2 + 2\langle \zeta - w, \zeta^* - w^*\rangle = 0.$$

Proof. (a) is immediate from Lemmas 20.3 and 20.2.
(b) Let $S_i := (T_i^{-1} - w)^{-1}$. From Lemma 15.3, S_1 and S_2 satisfy (20.4.1). The result follows from (a), with $\zeta := z + w$.
(c) Let $T_i := S_i - w^*/2$. From Lemma 15.4, T_1 and T_2 satisfy (20.4.2). The result follows from (b), with $\zeta^* := z^* + w^*$. ▌

Theorem 20.5. *Let E be reflexive, $S_1\colon E \mapsto 2^{E^*}$ and $S_2\colon E \mapsto 2^{E^*}$ be maximal monotone and*

$$\operatorname{dom}\chi_{S_1} \cap \operatorname{int}\operatorname{dom}\chi_{S_2} \neq \emptyset.$$

Then $S_1 + S_2$ is maximal monotone.

Proof. This is immediate from Lemma 20.4(c) and Theorem 10.3. ▌

We note from Lemma 10.5 that the conclusion of Lemma 20.4(a) is that $R(S_1 + S_2 + J) \ni 0$. If $w \in E$, define $J_w(x) := J(x - w) \quad (x \in E)$. Then the conclusion of Lemma 20.4(c) is that,

$$\text{for all } w \in E, \quad R(S_1 + S_2 + J_w) = E^*.$$

The simple bootstrapping procedure outlined in Lemma 20.4(b–c) enables us to avoid having to use a renorming theorem. (To be more precise, what we avoid is the use of the result that *E can be renormed so that if $S\colon E \mapsto 2^{E^*}$ is monotone and $S + J$ is surjective then S is maximal monotone.* See the comments preceding Theorem 10.7.) We also point out for the record that we do not use any fixed–point theorems either. Incidentally, Theorem 20.5 is only slightly stronger than Rockafellar's original result since, from Theorem 18.8,

$$\operatorname{int} D(S) = \operatorname{int}\operatorname{dom}\chi_S = \operatorname{int}\operatorname{dom}\psi_S.$$

21. The dom–dom constraint qualification

We shall say that S_1 and S_2 satisfy the "dom–dom constraint qualification" if

$$\operatorname{dom}\chi_{S_1} - \operatorname{dom}\chi_{S_2} \quad \text{is absorbing,} \tag{21.0.1}$$

and we shall prove in Theorem 21.3 that if E is reflexive and S_1 and S_2 are maximal monotone and satisfy the dom–dom constraint qualification then $S_1 + S_2$ is maximal monotone. We shall put this result into a historical perspective and discuss various other constraint qualifications in the remaining sections in this chapter.

Lemma 21.1, which assumes neither the reflexivity of E nor the maximality of S_1 or S_2, will be our first step towards Theorem 21.3. Lemma 21.1 is patterned after Lemma 20.3. It is, however, much harder since it uses the dom–dom lemma, Lemma 13.1 rather than the dom lemma, Lemma 12.2.

Lemma 21.1. *Let $S_1\colon E \mapsto 2^{E^*}$ and $S_2\colon E \mapsto 2^{E^*}$ be nontrivial and monotone, and satisfy the dom–dom constraint qualification. Then:*

$$\left.\begin{aligned} &\text{there exists } \gamma \geq 0 \text{ such that}\\ &\quad (\mu_1,\mu_2) \in \mathcal{CO}(S_1)\times\mathcal{CO}(S_2) \implies\\ &\qquad 2r(\mu_1+\mu_2) + 2\gamma\|p(\mu_2-\mu_1)\|\\ &\qquad\qquad + \|p(\mu_1)\|^2 + \|q(\mu_1+\mu_2)\|^2 \geq 0. \end{aligned}\right\} \tag{20.1.1}$$

Proof. To simplify the expressions in what follows, we write χ_i instead of χ_{S_i}. From the dom–dom lemma, Lemma 13.1, there exist $n \geq 1$ and $\eta \in (0,1]$ such that

$$\left.\begin{aligned} &w \in E \text{ and } \|w\| \leq \eta \implies\\ &\quad w \in \{x \in E\colon\ \chi_2(x) \leq n\} - \{x \in E\colon\ \chi_1(x) \vee \|x\| \leq n\}. \end{aligned}\right\} \tag{21.1.1}$$

We write $\gamma := 5n^2/\eta$, and we shall show that γ has the required property. So let $(\mu_1,\mu_2) \in \mathcal{CO}(S_1)\times\mathcal{CO}(S_2)$. If $\|q(\mu_2)\| \leq \gamma$, then the inequality in (20.1.1) follows from Corollary 9.2(b), so we only have to consider the other alternative, namely that $\|q(\mu_2)\| > \gamma$. We choose an element w of E such that

$$\|w\| \leq \eta \quad\text{and}\quad \langle w, q(\mu_2)\rangle \geq \gamma\eta = 5n^2, \tag{21.1.2}$$

From (21.1.1), there exist $x,\ y \in E$ such that

$$\chi_2(x) \leq n, \quad \chi_1(y) \leq n, \quad \|y\| \leq n \quad\text{and}\quad x - y = w.$$

Since $\chi_2(x) \leq n$,

$$r(\mu_2) + n(1 + \|p(\mu_2)\|) - \langle x, q(\mu_2)\rangle \geq 0.$$

Using the fact that $x = y + w$, we derive from (21.1.2) that

$$r(\mu_2) + n(1 + \|p(\mu_2)\|) - \langle y, q(\mu_2)\rangle - 5n^2 \geq 0. \tag{21.1.3}$$

Since $\chi_1(y) \leq n$,

$$r(\mu_1) + n(1 + \|p(\mu_1)\|) - \langle y, q(\mu_1)\rangle \geq 0. \tag{21.1.4}$$

Adding (21.1.3) and (21.1.4) and using the fact that

$$\|p(\mu_2)\| \leq \|p(\mu_2-\mu_1)\| + \|p(\mu_1)\|,$$

we obtain

$$r(\mu_1+\mu_2) + n\|p(\mu_2-\mu_1)\| + 2n\|p(\mu_1)\| - \langle y, q(\mu_1+\mu_2)\rangle - 3n^2 \geq 0.$$

Multiplying by 2 and using the fact that $\|y\| \leq n$,

$$2r(\mu_1+\mu_2) + 2n\|p(\mu_2-\mu_1)\| + 4n\|p(\mu_1)\| + 2n\|q(\mu_1+\mu_2)\| - 6n^2 \geq 0.$$

Since $4n\|p(\mu_1)\| \leq 4n^2 + \|p(\mu_1)\|^2$ and $2n\|q(\mu_1+\mu_2)\| \leq n^2 + \|q(\mu_1+\mu_2)\|^2$,

$$2r(\mu_1+\mu_2) + 2n\|p(\mu_2-\mu_1)\| + \|p(\mu_1)\|^2 + \|q(\mu_1+\mu_2)\|^2 - n^2 \geq 0.$$

The inequality in (20.1.1) follows from this since $n \leq \gamma$. ∎

Lemma 21.2 and Theorem 21.3 follow from Lemma 21.1 in exactly the same way that Lemma 20.4 and Theorem 20.5 followed from Lemma 20.3.

Lemma 21.2. *Let E be reflexive.*
(a) Let $S_1\colon E \mapsto 2^{E^*}$ and $S_2\colon E \mapsto 2^{E^*}$ be maximal monotone and satisfy the dom–dom constraint qualification. Then there exists $(z, z^*) \in G(S_1+S_2)$ such that

$$\|z\|^2 + \|z^*\|^2 + 2\langle z, z^*\rangle = 0.$$

(b) Let $T_1\colon E \mapsto 2^{E^*}$ and $T_2\colon E \mapsto 2^{E^*}$ be maximal monotone and satisfy the dom–dom constraint qualification. Suppose also that $w \in E$. Then there exists $(\zeta, z^*) \in G(T_1 + T_2)$ such that

$$\|\zeta - w\|^2 + \|z^*\|^2 + 2\langle \zeta - w, z^*\rangle = 0.$$

(c) Let $S_1\colon E \mapsto 2^{E^*}$ and $S_2\colon E \mapsto 2^{E^*}$ be maximal monotone and satisfy the dom–dom constraint qualification. Suppose also that $w \in E$ and $w^* \in E^*$. Then there exists $(\zeta, \zeta^*) \in G(S_1 + S_2)$ such that

$$\|\zeta - w\|^2 + \|\zeta^* - w^*\|^2 + 2\langle \zeta - w, \zeta^* - w^*\rangle = 0.$$

Theorem 21.3. *Let E be reflexive, $S_1\colon E \mapsto 2^{E^*}$ and $S_2\colon E \mapsto 2^{E^*}$ be maximal monotone and satisfy the dom–dom constraint qualification. Then $S_1 + S_2$ is maximal monotone.*

One can, in fact, obtain a necessary and sufficient condition for $S_1 + S_2$ to be maximal monotone. It is quite ugly, but at least it is an equivalence.

Theorem 21.4. *Let E be reflexive and $S_1\colon E \mapsto 2^{E^*}$ and $S_2\colon E \mapsto 2^{E^*}$ be maximal monotone. Then $S_1 + S_2$ is maximal monotone if, and only if, for all $w \in E$ and $w^* \in E^*$,*

$$\begin{aligned}
&\text{there exists } \gamma \geq 0 \text{ such that} \quad (\mu_1, \mu_2) \in \mathcal{CO}(S_1) \times \mathcal{CO}(S_2) \quad \Longrightarrow \\
&\quad 2r(\mu_1 + \mu_2) - 2\langle p(\mu_1), w^*\rangle - 2\langle w, q(\mu_1 + \mu_2)\rangle + 2\langle w, w^*\rangle \\
&\qquad + 2\gamma\|p(\mu_2 - \mu_1)\| + \|p(\mu_1) - w\|^2 + \|q(\mu_1 + \mu_2) - w^*\|^2 \geq 0.
\end{aligned}$$

Proof. It follows from Theorem 10.3 and the logic of Lemma 20.2 that S_1+S_2 is maximal monotone if, and only if, for all $w \in E$ and $w^* \in E^*$,

$$\begin{aligned}
&\text{there exist } x_1^*,\ x_2^* \in E^* \text{ and } z \in E \text{ such that} \\
&\quad (s_1, s_1^*) \in G(S_1) \text{ and } (s_2, s_2^*) \in G(S_2) \quad \Longrightarrow \\
&\qquad 2\langle s_1 - z, s_1^* - x_1^*\rangle + 2\langle s_2 - z, s_2^* - x_2^*\rangle \geq \\
&\qquad\qquad \|z - w\|^2 + \|x_1^* + x_2^* - w^*\|^2 + 2\langle z - w, x_1^* + x_2^* - w^*\rangle.
\end{aligned}$$

We now obtain the required result by adapting the proof of the γ–lemma, Lemma 20.1 (exercise!). ▌

Problem 21.5. Let E be reflexive, $S_1\colon E \mapsto 2^{E^*}$ and $S_2\colon E \mapsto 2^{E^*}$ be maximal monotone and

$$\operatorname{dom}\psi_{S_1} - \operatorname{dom}\psi_{S_2} \quad \text{be absorbing.}$$

Then is S_1+S_2 necessarily maximal monotone? Of course, this problem only makes sense after we have found solutions to Problem 15.5.

22. The six set and the nine set theorems for pairs of multifunctions

Let E be reflexive and $S_1\colon E \mapsto 2^{E^*}$ and $S_2\colon E \mapsto 2^{E^*}$ be maximal monotone. Theorem 22.1(c) and Theorem 22.2(a) give analogs for *pairs* of maximal monotone multifunctions for *reflexive* spaces of the results proved in Theorems 18.3 and 18.4 for *single* multifunctions on a *general* Banach space. Of course, these results are much "harder" than the previous results since they use Lemma 20.2. Our main result here is the "six set theorem for pairs", Theorem 22.1(c), in which we prove the identity of $\operatorname{int}(D(S_1)-D(S_2))$ with five other sets. In the "nine set theorem for pairs", Theorem 22.2(a), we prove the identity of $\overline{D(S_1)-D(S_2)}$ with eight other sets if $D(S_1)-D(S_2)$ is sufficiently fat. The results of this section appear in our paper [57].

Theorem 22.1. *Let E be reflexive and $S_1\colon E \mapsto 2^{E^*}$ and $S_2\colon E \mapsto 2^{E^*}$ be maximal monotone. Then:*
(a) $\operatorname{Sur}(\operatorname{dom}\chi_{S_1} - \operatorname{dom}\chi_{S_2}) \subset D(S_1) - D(S_2)$.
(b) $\operatorname{Sur}(\operatorname{dom}\chi_{S_1} - \operatorname{dom}\chi_{S_2}) \subset \operatorname{int}(D(S_1) - D(S_2))$.
(c) We have
$\operatorname{Int}(D(S_1)-D(S_2)) = \operatorname{int}(\operatorname{co}D(S_1)-\operatorname{co}D(S_2)) = \operatorname{int}(\operatorname{dom}\chi_{S_1} - \operatorname{dom}\chi_{S_2}) = \operatorname{sur}(D(S_1)-D(S_2)) = \operatorname{sur}(\operatorname{co}D(S_1)-\operatorname{co}D(S_2)) = \operatorname{sur}(\operatorname{dom}\chi_{S_1} - \operatorname{dom}\chi_{S_2})$.
(d) $\operatorname{Int}(D(S_1)-D(S_2))$ *is convex.*
(e) $\operatorname{Sur}(\operatorname{co}D(S_1)-\operatorname{co}D(S_2)) \subset \operatorname{int}(D(S_1)-D(S_2))$.

Proof. Let $x \in \operatorname{sur}(\operatorname{dom}\chi_{S_1} - \operatorname{dom}\chi_{S_2})$. Define $S_3\colon E \mapsto 2^{E^*}$ by $S_3 := (S_1^{-1} - x)^{-1}$. From Lemma 15.3,

$$D(S_3) = D(S_1) - x \quad \text{and} \quad \operatorname{dom}\chi_{S_3} = \operatorname{dom}\chi_{S_1} - x \tag{22.1.1}$$

thus, by hypothesis,

$$0 \in \operatorname{sur}(\operatorname{dom}\chi_{S_1} - x - \operatorname{dom}\chi_{S_2}) = \operatorname{sur}(\operatorname{dom}\chi_{S_3} - \operatorname{dom}\chi_{S_2}).$$

From (12.0.1), S_3 and S_2 satisfy the dom–dom constraint qualification hence, from Lemma 20.2,

$$\text{there exists } z \in D(S_3+S_2) = D(S_3)\cap D(S_2).$$

From (22.1.1), there exists $y \in D(S_1)$ such that $z = y - x$. But then

$$x = y - z \in D(S_1) - D(S_2).$$

This completes the proof of (a). However, it follows from the dom–dom corollary, Corollary 13.2 that $\operatorname{sur}(\operatorname{dom}\chi_{S_1} - \operatorname{dom}\chi_{S_2})$ is open, thus (b) is a consequence of (a). From the D–dom lemma, Lemma 15.2,

$$\operatorname{int}(D(S_1) - D(S_2)) \subset \operatorname{int}(\operatorname{co}D(S_1) - \operatorname{co}D(S_2)) \subset \operatorname{int}(\operatorname{dom}\chi_{S_1} - \operatorname{dom}\chi_{S_2})$$

and

$$\operatorname{sur}(D(S_1) - D(S_2)) \subset \operatorname{sur}(\operatorname{co}D(S_1) - \operatorname{co}D(S_2)) \subset \operatorname{sur}(\operatorname{dom}\chi_{S_1} - \operatorname{dom}\chi_{S_2}).$$

Since $\operatorname{int}(\ldots) \subset \operatorname{sur}(\ldots)$, (c) follows from (b). Finally, (d) and (e) are immediate from (c). ∎

Theorem 22.2. *Let E be reflexive, $S_1\colon E \mapsto 2^{E^*}$ and $S_2\colon E \mapsto 2^{E^*}$ be maximal monotone multifunctions and*

$$\operatorname{sur}(\operatorname{dom}\chi_{S_1} - \operatorname{dom}\chi_{S_2}) \neq \emptyset.$$

Then:

(a) $\overline{D(S_1) - D(S_2)} = \overline{\operatorname{co}D(S_1) - \operatorname{co}D(S_2)} = \overline{\operatorname{dom}\chi_{S_1} - \operatorname{dom}\chi_{S_2}} =$
$\overline{\operatorname{int}(D(S_1) - D(S_2))} = \overline{\operatorname{int}(\operatorname{co}D(S_1) - \operatorname{co}D(S_2))} = \overline{\operatorname{int}(\operatorname{dom}\chi_{S_1} - \operatorname{dom}\chi_{S_2})} =$
$\overline{\operatorname{sur}(D(S_1) - D(S_2))} = \overline{\operatorname{sur}(\operatorname{co}D(S_1) - \operatorname{co}D(S_2))} = \overline{\operatorname{sur}(\operatorname{dom}\chi_{S_1} - \operatorname{dom}\chi_{S_2})}$.

(b) $\overline{D(S_1) - D(S_2)}$ *is convex.*

Proof. From the D–dom lemma, Lemma 15.2,

$$\overline{D(S_1) - D(S_2)} \subset \overline{\operatorname{co}D(S_1) - \operatorname{co}D(S_2)} \subset \overline{\operatorname{dom}\chi_{S_1} - \operatorname{dom}\chi_{S_2}}.$$

Clearly,

$$\overline{\operatorname{int}(D(S_1) - D(S_2))} \subset \overline{D(S_1) - D(S_2)}.$$

From the dom–dom corollary, Corollary 13.2, $\operatorname{int}(\operatorname{dom}\chi_{S_1} - \operatorname{dom}\chi_{S_2}) \neq \emptyset$ hence (see, for instance, Kelly–Namioka, [28], 13.1(i), p. 100–111)

$$\overline{\operatorname{dom}\chi_{S_1} - \operatorname{dom}\chi_{S_2}} = \overline{\operatorname{int}(\operatorname{dom}\chi_{S_1} - \operatorname{dom}\chi_{S_2})}.$$

(a) now follows from Theorem 22.1(c), and (b) is an immediate consequence of (a). ∎

23. The equivalence of six constraint qualifications — twice

In this section, we discuss various constraint qualifications that have been shown to imply the maximal monotonicity of $S_1 + S_2$ when E is reflexive and S_1 and S_2 are maximal monotone. As we have already observed, the original result due to Rockafellar was that this is the case if

$$D(S_1) \cap \operatorname{int} D(S_2) \neq \emptyset. \tag{20.0.1}$$

Attouch–Riahi–Théra proved (see [2], Théorème 4) that this is also the case if

$$D(S_1) - D(S_2) \quad \text{is absorbing,} \tag{23.1.2}$$

and Chu proved (this is contained in [16], Corollary 3.5) that this is also the case if

$$\operatorname{co}D(S_1) - \operatorname{co}D(S_2) \quad \text{is a neighborhood of } 0. \tag{23.1.5}$$

Finally, we have proved in Theorem 21.3 that this is also the case if

$$\operatorname{dom}\chi_{S_1} - \operatorname{dom}\chi_{S_2} \quad \text{is absorbing.} \tag{23.1.6}$$

It is clear from the D–dom lemma, Lemma 15.2 that Theorem 21.3 is a formal generalization of the preceding three results. We shall show in Theorem 23.1 that (23.1.2), (23.1.5) and (23.1.6) are, in fact, equivalent. This equivalence is not trivial, relying as it does on Theorem 22.1(a), which relies on Lemma 20.2. We point out that the analysis in [43], [2] and [16] implicitly uses the renorming of E (see the remarks preceding Theorem 10.7).

Theorem 23.1. *Let E be reflexive and $S_1\colon E \mapsto 2^{E^*}$ and $S_2\colon E \mapsto 2^{E^*}$ be maximal monotone. Then the six conditions below are equivalent:*

$$D(S_1) - D(S_2) \quad \textit{is a neighborhood of } 0 \tag{23.1.1}$$

$$D(S_1) - D(S_2) \quad \textit{is absorbing} \tag{23.1.2}$$

$$D(S_1) - D(S_2) \quad \textit{surrounds } 0 \tag{23.1.3}$$

$$\operatorname{co}D(S_1) - \operatorname{co}D(S_2) \quad \textit{is absorbing} \tag{23.1.4}$$

$$\operatorname{co}D(S_1) - \operatorname{co}D(S_2) \quad \textit{is a neighborhood of } 0 \tag{23.1.5}$$

$$\operatorname{dom}\chi_{S_1} - \operatorname{dom}\chi_{S_2} \quad \textit{is absorbing.} \tag{23.1.6}$$

Proof. Using (12.0.1), the equivalence of (23.1.1), (23.1.3), (23.1.4), (23.1.5) and (23.1.6) follows by putting "$0 \in$" in front of the first, fourth, fifth, second and sixth sets, respectively, of Theorem 22.1(c). (23.1.2) is intermediate between (23.1.1) and (23.1.3). ▮

In fact, there are also "subspace" versions of the above constraint qualifications that are known to be sufficient. Attouch-Riahi-Théra have proved (see [2], Corollaire 1) that (23.1.2) can be weakened to

$$\bigcup_{\lambda>0} \lambda[D(S_1) - D(S_2)] = \overline{\mathrm{lin}(D(S_1) - D(S_2))}, \tag{23.2.2}$$

while Chu has proved (see [16], Corollary 3.5) that (20.0.1) can be weakened to

$$\left.\begin{array}{l}\mathrm{co}\, D(S_1) - \mathrm{co}\, D(S_2) \quad \text{is a neighborhood of 0 in} \\ \overline{\mathrm{lin}(D(S_1) - D(S_2))}.\end{array}\right\} \tag{23.2.3}$$

Theorem 23.2 generalizes both of these results. Theorem 23.2 appears in Simons, [57], Theorem 49. The proof given here is simplified considerably by the analysis in Section 16.

Theorem 23.2. *Let E be a reflexive Banach space and $S_1\colon E \mapsto 2^{E^*}$ and $S_2\colon E \mapsto 2^{E^*}$ be maximal monotone. Then the conditions (23.2.1) — (23.2.6) are equivalent, and imply that $S_1 + S_2$ is maximal monotone:*

$$D(S_1) - D(S_2) \quad \text{is a neighborhood of 0 in} \quad \overline{\mathrm{lin}(D(S_1) - D(S_2))} \tag{23.2.1}$$

$$\bigcup_{\lambda>0} \lambda[D(S_1) - D(S_2)] = \overline{\mathrm{lin}(D(S_1) - D(S_2))} \tag{23.2.2}$$

$$\left.\begin{array}{l}\mathrm{co} D(S_1) - \mathrm{co} D(S_2) \quad \text{is a neighborhood of 0 in} \\ \overline{\mathrm{lin}(D(S_1) - D(S_2))}\end{array}\right\} \tag{23.2.3}$$

$$\bigcup_{\lambda>0} \lambda[\mathrm{co}\, D(S_1) - \mathrm{co}\, D(S_2)] = \overline{\mathrm{lin}(D(S_1) - D(S_2))} \tag{23.2.4}$$

$$\left.\begin{array}{l}\mathrm{dom}\, \chi_{S_1} - \mathrm{dom}\, \chi_{S_2} \quad \text{is a neighborhood of 0 in} \\ \overline{\mathrm{lin}(\mathrm{dom}\, \chi_{S_1} - \mathrm{dom}\, \chi_{S_2})}\end{array}\right\} \tag{23.2.5}$$

$$\bigcup_{\lambda>0} \lambda[\mathrm{dom}\, \chi_{S_1} - \mathrm{dom}\, \chi_{S_2}] = \overline{\mathrm{lin}(\mathrm{dom}\, \chi_{S_1} - \mathrm{dom}\, \chi_{S_2})}. \tag{23.2.6}$$

Proof. Clearly (23.2.1) $\Longrightarrow$ (23.2.2) $\Longrightarrow$ (23.2.4), (23.2.1) $\Longrightarrow$ (23.2.3) $\Longrightarrow$ (23.2.4) and (23.2.5) $\Longrightarrow$ (23.2.6). Further, it follows from the D–dom lemma, Lemma 15.2, and Theorem 16.8 that (23.2.1) $\Longrightarrow$ (23.2.5) and (23.2.4) $\Longrightarrow$ (23.2.6). We shall prove that (23.2.6) $\Longrightarrow$ (23.2.1), and that if (23.2.6) is satisfied then $S_1 + S_2$ is maximal monotone.

((23.2.6)$\Longrightarrow$(23.2.1)) Suppose that (23.2.6) is satisfied, and let

$$F := \overline{\operatorname{lin}(\operatorname{dom}\chi_{S_1} - \operatorname{dom}\chi_{S_2})}.$$

Since $0 \in F$, it follows from (23.2.6) that $0 \in \operatorname{dom}\chi_{S_1} - \operatorname{dom}\chi_{S_2}$, hence there exists $w \in \operatorname{dom}\chi_{S_1} \cap \operatorname{dom}\chi_{S_2}$. For $i = 1,\ 2$, let $T_i := (S_i^{-1} - w)^{-1}$. We have from the D–dom lemma, Lemma 15.2, that

$$D(T_1) = D(S_1) - w \subset \operatorname{dom}\chi_{S_1} - w \subset \operatorname{dom}\chi_{S_1} - \operatorname{dom}\chi_{S_2} \subset F \qquad (23.2.7)$$

and

$$D(T_2) = D(S_2) - w \subset \operatorname{dom}\chi_{S_2} - w \subset \operatorname{dom}\chi_{S_2} - \operatorname{dom}\chi_{S_1} \subset F. \qquad (23.2.8)$$

Clearly, T_1 and T_2 are maximal monotone thus, from Lemmas 15.3, 16.5 and Theorem 16.10(d),

$$\operatorname{dom}\chi_{(T_{1F})} = \operatorname{dom}\chi_{T_1} = \operatorname{dom}\chi_{S_1} - w$$

and

$$\operatorname{dom}\chi_{(T_{2F})} = \operatorname{dom}\chi_{T_2} = \operatorname{dom}\chi_{S_2} - w.$$

Substituting the above into (23.2.6),

$$\bigcup_{\lambda>0} \lambda\big[\operatorname{dom}\chi_{(T_{1F})} - \operatorname{dom}\chi_{(T_{2F})}\big] = F.$$

From Theorem 16.10(e)($\Longrightarrow$), T_{1F} and T_{2F} are maximal monotone hence, from Theorem 21.3 and Theorem 23.1 with E replaced by F,

$$T_{1F} + T_{2F} \text{ is maximal monotone} \qquad (23.2.9)$$

and

$$D(T_{1F}) - D(T_{2F}) \quad \text{is a neighborhood of } 0 \text{ in } F. \qquad (23.2.10)$$

Now, from (23.2.7), (23.2.8) and Theorem 16.10(b),

$$D(S_1) - D(S_2) = D(T_1) - D(T_2) = D(T_{1F}) - D(T_{2F}),$$

and so (23.2.1) follows from (23.2.10). This completes the proof that (23.2.6) $\Longrightarrow$ (23.2.1), and also completes the proof of the equivalence of the six conditions.

If any of the six conditions is satisfied then so also is (23.2.9). Since $(T_1 + T_2)_F = T_{1F} + T_{2F}$, $(T_1 + T_2)_F$ is maximal monotone. From Lemma 16.5, T_1 and T_2 are F-saturated, hence so also is $T_1 + T_2$. We now deduce from Theorem 16.10(e)($\Longleftarrow$) that $T_1 + T_2$ is maximal monotone, from which it follows that $S_1 + S_2$ is maximal monotone too. ∎

24. The Brézis–Crandall–Pazy condition

At this point, we shall step back a little and try to give an overview of the results that we have established so far on the problem of the maximal monotonicity of the sum of two maximal monotone multifunctions on a reflexive Banach space. In the γ–lemma, Lemma 20.1, we established the equivalence of a number of conditions which will lead to the maximal monotonicity of the sum provided that the bootstrapping steps outlined in Lemma 20.4(b–c) are valid. Further, using Baire's theorem, we proved in Lemma 21.1 that if the dom–dom constraint qualification is satisfied, then so also are the conditions of the γ–lemma.

On the other hand, we will prove in Lemma 24.1 below that, for any maximal monotone multifunctions S_1 and S_2 such that

$$\operatorname{co} D(S_1) \cap \operatorname{co} D(S_2) \neq \emptyset \tag{24.0.1}$$

without further restriction, (24.1.4) is satisfied. (The condition (24.0.1) is clearly satisfied if $S_1 + S_2$ is maximal monotone.)

(24.1.4) is manifestly close to the γ–condition, (20.1.1); in this section, we discuss a totally different kind of constraint qualification that does not depend on Baire's theorem and enables us to deduce the γ–condition from (24.1.4). This condition is automatically satisfied if S_1 and S_2 satisfy the *Brézis–Crandall–Pazy condition*, which means that there exist increasing functions k: $[0,\infty) \mapsto [0,1)$ and C: $[0,\infty) \mapsto [0,\infty)$ such that, writing $|S_k x_1| := \inf \|S_k x_1\|$,

$$\left.\begin{array}{ll} \emptyset \neq D(S_1) \subset D(S_2) \quad \text{and} & \\ x_1 \in D(S_1) \quad \Longrightarrow & |S_2 x_1| \leq k(\|x_1\|)|S_1 x_1| + C(\|x_1\|). \end{array}\right\} \tag{24.0.2}$$

The Brézis–Crandall–Pazy condition condition can be thought of as a *perturbation condition*, and has found application to partial differential equations. We refer the reader to the original paper by Brézis, Crandall and Pazy, [12], for more details. The most general result in this section on the maximal monotonicity of $S_1 + S_2$ is Theorem 24.3. We show in Theorem 24.4 how to deduce from it the main result of [12].

It is very important in Lemma 24.1 that R be independent of n.

Lemma 24.1. *Let E be reflexive, S_1: $E \mapsto 2^{E^*}$ and S_2: $E \mapsto 2^{E^*}$ be maximal monotone and* $\operatorname{co} D(S_1) \cap \operatorname{co} D(S_2) \neq \emptyset$. *Then there exists $R \geq 0$ such that, for all $n \geq 1$, there exist $(x_1, x_1^*) \in G(S_1)$ and $(x_2, x_2^*) \in G(S_2)$ such that*

$$\|x_1\|^2 + 2\langle x_1, x_1^* + x_2^* \rangle + \|x_1^* + x_2^*\|^2 = 0, \tag{24.1.1}$$

$$\langle x_1 - x_2, x_2^* \rangle = \|x_1 - x_2\| \|x_2^*\|, \tag{24.1.2}$$

$$\|x_1\|^2 + \|x_1^* + x_2^*\|^2 \le R^2, \tag{24.1.3}$$

and

$$\left.\begin{aligned} &(\mu_1, \mu_2) \in \mathcal{CO}(S_1) \times \mathcal{CO}(S_2) \implies \\ &\quad 2r(\mu_1 + \mu_2) + 2\|x_2^*\|\|p(\mu_2 - \mu_1)\| \\ &\qquad + \|p(\mu_1)\|^2 + \|q(\mu_1 + \mu_2)\|^2 + 2R\|q(\mu_2)\|/n \ge 0. \end{aligned}\right\} \tag{24.1.4}$$

Proof. Since $\operatorname{co} D(S_1) \cap \operatorname{co} D(S_2) \neq \emptyset$, we can fix $(\nu_1, \nu_2) \in \mathcal{CO}(S_1) \times \mathcal{CO}(S_2)$ so that $p(\nu_1) = p(\nu_2)$. We then define

$$\left.\begin{aligned} R := &\sqrt{\|q(\nu_1 + \nu_2)\|^2 + \|q(\nu_2)\|^2 + \|p(\nu_1)\|^2} \\ &+ \sqrt{2r(\nu_1 + \nu_2) + \|q(\nu_1 + \nu_2)\|^2 + \|q(\nu_2)\|^2 + \|p(\nu_1)\|^2}. \end{aligned}\right\} \tag{24.1.5}$$

(The first square root above is obviously real. The second one is also real since, from the *pqr*–lemma, Lemma 9.1,

$$2r(\nu_1 + \nu_2) \ge 2\langle p(\nu_1), q(\nu_1)\rangle + 2\langle p(\nu_2), q(\nu_2)\rangle = 2\langle p(\nu_1), q(\nu_1 + \nu_2)\rangle.)$$

Since $2\|p(\mu_2 - \mu_1)\|\|q(\mu_2)\| \le n^2\|p(\mu_2 - \mu_1)\|^2 + \|q(\mu_2)\|/n^2$, it follows from Corollary 9.2(b) that, for all $(\mu_1, \mu_2) \in \mathcal{CO}(S_1) \times \mathcal{CO}(S_2)$,

$$2r(\mu_1 + \mu_2) + n^2\|p(\mu_2 - \mu_1)\|^2 + \|q(\mu_2)\|/n^2 + \|p(\mu_1)\|^2 + \|q(\mu_1 + \mu_2)\|^2 \ge 0.$$

We write $F := E \times E \times E^* \times E^*$ with

$$\|(x_1, x_2, x_1^*, x_2^*)\| := \sqrt{\|x_1\|^2 + \|x_2\|^2 + \|x_1^*\|^2 + \|x_2^*\|^2}$$

and, for all $(\mu_1, \mu_2) \in \mathcal{CO}(S_1) \times \mathcal{CO}(S_2)$,

$$f(\mu_1, \mu_2) := 2r(\mu_1 + \mu_2)$$

and

$$g(\mu_1, \mu_2) := (p(\mu_1), np(\mu_2 - \mu_1), q(\mu_1 + \mu_2), q(\mu_2)/n).$$

It follows from the *fg*–theorem, Theorem 7.2, that there exists $(z_1, z_2, z_1^*, z_2^*) \in F$ such that, for all $(\mu_1, \mu_2) \in \mathcal{CO}(S_1) \times \mathcal{CO}(S_2)$,

$$\left.\begin{aligned} 2r(\mu_1 + \mu_2) &- 2\langle p(\mu_1), z_1^*\rangle - 2\langle np(\mu_2 - \mu_1), z_2^*\rangle \\ &- 2\langle z_1, q(\mu_1 + \mu_2)\rangle - 2\langle z_2, q(\mu_2)/n\rangle \\ &\ge \|z_1\|^2 + \|z_2\|^2 + \|z_1^*\|^2 + \|z_2^*\|^2, \end{aligned}\right\} \tag{24.1.6}$$

that is to say,

$$\begin{aligned} 2r(\mu_1 + \mu_2) &- 2\langle p(\mu_1), z_1^* - nz_2^*\rangle - 2\langle p(\mu_2), nz_2^*\rangle \\ &- 2\langle z_1, q(\mu_1)\rangle - 2\langle z_1 + z_2/n, q(\mu_2)\rangle \\ &\ge \|z_1\|^2 + \|z_2\|^2 + \|z_1^*\|^2 + \|z_2^*\|^2. \end{aligned}$$

If we now restrict μ_1 and μ_2 to the values $\delta_{(s_1,s_1^*)}$ and $\delta_{(s_2,s_2^*)}$ we obtain that, for all $(s_1, s_1^*) \in G(S_1)$ and $(s_2, s_2^*) \in G(S_2)$,

$$\begin{aligned}2\langle s_1, s_1^*\rangle + 2\langle s_2, s_2^*\rangle - 2\langle s_1, z_1^* - nz_2^*\rangle - 2\langle s_2, nz_2^*\rangle \\ - 2\langle z_1, s_1^*\rangle - 2\langle z_1 + z_2/n, s_2^*\rangle \\ \geq \|z_1\|^2 + \|z_2\|^2 + \|z_1^*\|^2 + \|z_2^*\|^2.\end{aligned}$$

After some simple manipulations, this can be rewritten

$$\begin{aligned}2\langle s_1 - z_1, s_1^* - (z_1^* - nz_2^*)\rangle + 2\langle s_2 - (z_1 + z_2/n), s_2^* - nz_2^*\rangle \\ \geq \|z_1\|^2 + 2\langle z_1, z_1^*\rangle + \|z_1^*\|^2 + \|z_2\|^2 + 2\langle z_2, z_2^*\rangle + \|z_2^*\|^2.\end{aligned}$$

Arguing exactly as in Lemma 20.2, we obtain from this that $(z_1, z_1^* - nz_2^*) \in G(S_1)$, $(z_1 + z_2/n, nz_2^*) \in G(S_2)$,

$$\|z_1\|^2 + 2\langle z_1, z_1^*\rangle + \|z_1^*\|^2 = 0 \tag{24.1.7}$$

and $\|z_2\|^2 + 2\langle z_2, z_2^*\rangle + \|z_2^*\|^2 = 0$, from which

$$\langle z_2, z_2^*\rangle = -\|z_2\|\|z_2^*\|. \tag{24.1.8}$$

We now put

$$(x_1, x_1^*) := (z_1, z_1^* - nz_2^*) \in G(S_1)$$

and

$$(x_2, x_2^*) := (z_1 + z_2/n, nz_2^*) \in G(S_2).$$

Then $(z_1, z_1^*) = (x_1, x_1^* + x_2^*)$ and $(z_2, z_2^*) = (n(x_2 - x_1), x_2^*/n)$, and so (24.1.1) and (24.1.2) follow from (24.1.7) and (24.1.8). We thus obtain from (24.1.6) and the fact that $p(\nu_1) = p(\nu_2)$ that

$$\begin{aligned}\|z_1\|^2 + \|z_2\|^2 + \|z_1^*\|^2 &\leq 2r(\nu_1 + \nu_2) - 2\langle p(\nu_1), z_1^*\rangle \\ &\qquad - 2\langle z_1, q(\nu_1 + \nu_2)\rangle - 2\langle z_2, q(\nu_2)/n\rangle \\ &\leq 2r(\nu_1 + \nu_2) + 2\|z_1\|\|q(\nu_1 + \nu_2)\| \\ &\qquad + 2\|z_2\|\|q(\nu_2)\| + 2\|p(\nu_1)\|\|z_1^*\|.\end{aligned}$$

Consequently (compare with the proof of Lemma 10.9),

$$\begin{aligned}(\|z_1\| - \|q(\nu_1 + \nu_2)\|)^2 + (\|z_2\| - \|q(\nu_2)\|)^2 + (\|z_1^*\| - \|p(\nu_1)\|)^2 \\ \leq 2r(\nu_1 + \nu_2) + \|q(\nu_1 + \nu_2)\|^2 + \|q(\nu_2)\|^2 + \|p(\nu_1)\|^2,\end{aligned}$$

and it now follows from the triangle inequality in $\mathbb{R}^3$ and (24.1.5) that

$$\|z_1\|^2 + \|z_2\|^2 + \|z_1^*\|^2 \leq R^2. \tag{24.1.9}$$

We now obtain (24.1.3) since $z_1 = x_1$ and $z_1^* = x_1^* + x_2^*$.

Finally, let $(\mu_1, \mu_2) \in \mathcal{CO}(S_1) \times \mathcal{CO}(S_2)$. From the perfect square trick, Lemma 7.1,

$$\|z_1^*\|^2 \geq -2\langle p(\mu_1), z_1^* \rangle - \|p(\mu_1)\|^2$$

and

$$\|z_1\|^2 \geq -2\langle z_1, q(\mu_1 + \mu_2) \rangle - \|q(\mu_1 + \mu_2)\|^2$$

thus, from (24.1.6),

$$\left. \begin{aligned} 2r(\mu_1 + \mu_2) + \|p(\mu_1)\|^2 - 2\langle np(\mu_2 - \mu_1), z_2^* \rangle \\ + \|q(\mu_1 + \mu_2)\|^2 - 2\langle z_2, q(\mu_2)/n \rangle \geq \|z_2\|^2 + \|z_2^*\|^2 \geq 0. \end{aligned} \right\} \quad (24.1.10)$$

Using (24.1.9), we have $\|z_2\| \leq R$, and (24.1.4) then follows by substituting this into (24.1.10) and using the fact that $nz_2^* = x_2^*$. ∎

The bootstrapping argument in Lemma 24.2 is structurally similar to that in Lemmas 20.4 and 21.2, except that the nature of the constraint qualification here seems to force a different argument in the passage from (b) to (c).

Lemma 24.2. *Let E be reflexive.*
(a) *Let $S_1\colon E \mapsto 2^{E^*}$ and $S_2\colon E \mapsto 2^{E^*}$ be maximal monotone. Suppose that*

$$\operatorname{co} D(S_1) \cap \operatorname{co} D(S_2) \neq \emptyset$$

and that there exists an increasing function $j\colon [0, \infty) \mapsto [0, \infty)$ such that

$$(x_1, x_1^*) \in G(S_1),\ (x_2, x_2^*) \in G(S_2),\ x_1 \neq x_2 \text{ and } \langle x_1 - x_2, x_2^* \rangle = \|x_1 - x_2\| \|x_2^*\|$$
$$\Longrightarrow$$
$$\|x_2^*\| \leq j\left(\sqrt{\|x_1\|^2 + \|x_1^* + x_2^*\|^2}\right).$$

Then there exists $(z, z^*) \in G(S_1 + S_2)$ such that

$$\|z\|^2 + \|z^*\|^2 + 2\langle z, z^* \rangle = 0.$$

(b) *Let $T_1\colon E \mapsto 2^{E^*}$ and $T_2\colon E \mapsto 2^{E^*}$ be maximal monotone. Suppose that*

$$\operatorname{co} D(T_1) \cap \operatorname{co} D(T_2) \neq \emptyset$$

and that there exists an increasing function $k\colon [0, \infty) \mapsto [0, \infty)$ such that

$$(y_1, y_1^*) \in G(T_1),\ (y_2, y_2^*) \in G(T_2),\ y_1 \neq y_2 \text{ and } \langle y_1 - y_2, y_2^* \rangle = \|y_1 - y_2\| \|y_2^*\|$$
$$\Longrightarrow$$
$$\|y_2^*\| \leq k\left(\sqrt{\|y_1\|^2 + \|y_1^* + y_2^*\|^2}\right).$$

Suppose also that $w \in E$. Then there exists $(\zeta, z^*) \in G(T_1 + T_2)$ such that

$$\|\zeta - w\|^2 + \|z^*\|^2 + 2\langle \zeta - w, z^* \rangle = 0.$$

(c) *Let* $S_1\colon E \mapsto 2^{E^*}$ *and* $S_2\colon E \mapsto 2^{E^*}$ *be maximal monotone. Suppose that*

$$\operatorname{co} D(S_1) \cap \operatorname{co} D(S_2) \neq \emptyset$$

and that there exists an increasing function $j\colon [0,\infty) \mapsto [0,\infty)$ *such that*

$$(x_1, x_1^*) \in G(S_1),\ (x_2, x_2^*) \in G(S_2),\ x_1 \neq x_2 \text{ and } \langle x_1 - x_2, x_2^* \rangle = \|x_1 - x_2\| \|x_2^*\|$$

$$\Longrightarrow$$

$$\|x_2^*\| \le j\left(\sqrt{\|x_1\|^2 + \|x_1^* + x_2^*\|^2}\right).$$

Suppose also that $w \in E$ *and* $w^* \in E^*$. *Then there exists* $(\zeta, \zeta^*) \in G(S_1 + S_2)$ *such that*

$$\|\zeta - w\|^2 + \|\zeta^* - w^*\|^2 + 2\langle \zeta - w, \zeta^* - w^* \rangle = 0.$$

Proof. (a) Let R be as in Lemma 24.1 and $n \ge 1$. From Lemma 24.1, there exist $(x_1, x_1^*) \in G(S_1)$ and $(x_2, x_2^*) \in G(S_2)$ such that (24.1.1) — (24.1.4) are satisfied. If $x_1 = x_2$ then the result follows from (24.1.1) with $z := x_1 = x_2$ and $z^* := x_1^* + x_2^*$. If $x_1 \neq x_2$ then, from (24.1.2), we can use the hypothesis for j and so, from (24.1.3),

$$\|x_2^*\| \le j\left(\sqrt{\|x_1\|^2 + \|x_1^* + x_2^*\|^2}\right) \le j(R).$$

From (24.1.4), for all $(\mu_1, \mu_2) \in \mathcal{CO}(S_1) \times \mathcal{CO}(S_2)$,

$$2r(\mu_1 + \mu_2) + 2j(R)\|p(\mu_2 - \mu_1)\| + \|p(\mu_1)\|^2 + \|q(\mu_1 + \mu_2)\|^2 + 2R\|q(\mu_2)\|/n \ge 0.$$

Letting $n \to \infty$ in this, we have proved that the γ–condition, (20.1.1), is satisfied with $\gamma := j(R)$. The result now follows from Lemma 20.2.

(b) Let $S_1 := (T_1^{-1} - w)^{-1}$ and $S_2 := (T_2^{-1} - w)^{-1}$. If $(x_1, x_1^*) \in G(S_1)$, $(x_2, x_2^*) \in G(S_2)$, $x_1 \neq x_2$ and $\langle x_1 - x_2, x_2^* \rangle = \|x_1 - x_2\| \|x_2^*\|$ then we put $(y_1, y_1^*) := (x_1 + w, x_1^*) \in G(T_1)$ and $(y_2, y_2^*) := (x_2 + w, x_2^*) \in G(T_2)$. Clearly $y_1 \neq y_2$ and $\langle y_1 - y_2, y_2^* \rangle = \|y_1 - y_2\| \|y_2^*\|$. By hypothesis,

$$\begin{aligned} \|x_2^*\| = \|y_2^*\| &\le k\left(\sqrt{\|y_1\|^2 + \|y_1^* + y_2^*\|^2}\right) \\ &= k\left(\sqrt{\|x_1 + w\|^2 + \|x_1^* + x_2^*\|^2}\right) \\ &\le k\left(\sqrt{(\|x_1\| + \|w\|)^2 + \|x_1^* + x_2^*\|^2}\right). \end{aligned}$$

Using the triangle inequality in $\mathbb{R}^2$, we obtain

$$\|x_2^*\| \le k\left(\sqrt{\|x_1\|^2 + \|x_1^* + x_2^*\|^2} + \|w\|\right).$$

(b) now follows from (a) with $j(\rho) := k(\rho + \|w\|)$ and $\zeta := z + w$.

(c) Let $T_1 := S_1 - w^*$ and $T_2 := S_2$. If $(y_1, y_1^*) \in G(T_1)$, $(y_2, y_2^*) \in G(T_2)$, $y_1 \neq y_2$ and $\langle y_1 - y_2, y_2^* \rangle = \|y_1 - y_2\| \|y_2^*\|$ then we put $(x_1, x_1^*) := (y_1, y_1^* + w^*) \in G(S_1)$ and $(x_2, x_2^*) := (y_2, y_2^*) \in G(S_2)$. Clearly $x_1 \neq x_2$ and $\langle x_1 - x_2, x_2^* \rangle = \|x_1 - x_2\| \|x_2^*\|$. By hypothesis,

$$\begin{aligned}\|y_2^*\| = \|x_2^*\| &\le j\left(\sqrt{\|x_1\|^2 + \|x_1^* + x_2^*\|^2}\right)\\ &= j\left(\sqrt{\|y_1\|^2 + \|y_1^* + y_2^* + w^*\|^2}\right)\\ &\le j\left(\sqrt{\|y_1\|^2 + (\|y_1^* + y_2^*\| + \|w^*\|)^2}\right).\end{aligned}$$

Using the triangle inequality in $\mathbb{R}^2$, we obtain

$$\|y_2^*\| \le j\left(\sqrt{\|y_1\|^2 + \|y_1^* + y_2^*\|^2} + \|w^*\|\right).$$

(c) now follows from (b) with $k(\rho) := j(\rho + \|w^*\|)$ and $\zeta^* := z^* + w^*$. ∎

Theorem 24.3. *Let E be reflexive and $S_1\colon E \mapsto 2^{E^*}$ and $S_2\colon E \mapsto 2^{E^*}$ be maximal monotone. Suppose that*

$$\operatorname{co} D(S_1) \cap \operatorname{co} D(S_2) \neq \emptyset$$

and that there exists an increasing function $j\colon [0,\infty) \mapsto [0,\infty)$ such that

$$(x_1, x_1^*) \in G(S_1),\ (x_2, x_2^*) \in G(S_2),\ x_1 \neq x_2 \text{ and } \langle x_1 - x_2, x_2^* \rangle = \|x_1 - x_2\| \|x_2^*\|$$
$$\Longrightarrow$$
$$\|x_2^*\| \le j\left(\sqrt{\|x_1\|^2 + \|x_1^* + x_2^*\|^2}\right).$$

Then $S_1 + S_2$ is maximal monotone.

Proof. This is immediate from Lemma 24.2(c) and Theorem 10.3. ∎

Theorem 24.4. *Let E be reflexive and $S_1\colon E \mapsto 2^{E^*}$ and $S_2\colon E \mapsto 2^{E^*}$ be maximal monotone and satisfy the Brézis–Crandall–Pazy condition (24.0.2). Then $S_1 + S_2$ is maximal monotone.*

Proof. Suppose that $(x_1, x_1^*) \in G(S_1)$, $(x_2, x_2^*) \in G(S_2)$, $x_1 \neq x_2$ and

$$\langle x_1 - x_2, x_2^* \rangle = \|x_1 - x_2\| \|x_2^*\|. \tag{24.4.1}$$

Let y_2^* be an arbitrary element of $S_2 x_1$. Then, using (24.4.1), the monotonicity of S_2 and the fact that $(x_2, x_2^*) \in G(S_2)$,

$$\|x_1 - x_2\| \|x_2^*\| = \langle x_1 - x_2, x_2^* \rangle \le \langle x_1 - x_2, y_2^* \rangle \le \|x_1 - x_2\| \|y_2^*\|,$$

Dividing by $\|x_1 - x_2\|$, $\|x_2^*\| \le \|y_2^*\|$. Taking the infimum over $y_2^* \in S_2x_1$ gives $\|x_2^*\| \le |S_2x_1|$ and thus, from (24.0.2) with $\rho := \sqrt{\|x_1\|^2 + \|x_1^* + x_2^*\|^2}$,

$$\begin{aligned}
\|x_2^*\| &\le k(\|x_1\|)|S_1x_1| + C(\|x_1\|)\\
&\le k(\rho)\|x_1^*\| + C(\rho)\\
&\le k(\rho)(\|x_2^*\| + \|x_1^* + x_2^*\|) + C(\rho)\\
&\le k(\rho)(\|x_2^*\| + \rho) + C(\rho).
\end{aligned}$$

Since $\operatorname{co} D(S_1) \cap \operatorname{co} D(S_2) = \operatorname{co} D(S_1) \neq \emptyset$, the result now follows from Theorem 24.3 with

$$j(\rho) := \frac{\rho k(\rho) + C(\rho)}{1 - k(\rho)}. \blacksquare$$

Remark 24.5. We emphasize that, unlike the analysis in [12], we do not use any renorming or fixed–point theorems in any of the above results. As was pointed out in [12], one of the limitations of Rockafellar's original constraint qualification (20.0.1) was that it required $D(S_2)$ to have a nonempty interior. Of course, this is no longer true of the other constraint qualifications discussed in Theorems 23.1 and 23.2. However, even these conditions involve a "fatness" condition of a difference set relative to an enveloping closed subspace. On the other hand, Theorem 24.4 has the different limitation that $D(S_1) \subset D(S_2)$. Theorem 24.3 does not have either of the above limitations, though we do not know if it has any practical applications other than those that can be obtained from Theorem 24.4.

VI. Special maximal monotone multifunctions

25. Subclasses of the maximal monotone multifunctions

In recent years, many subclasses of the class of maximal monotone multifunctions have been introduced. In this section we introduce those that are "of type (D)", those that are "of type (FP)", those that are "of type (FPV)", those that are "of type (NI)", those that are "strongly maximal monotone", and those that are "of type (ANA)".

We first define multifunctions of type (D). These were essentially introduced by Gossez in [26], Lemme 2.1, p. 375 — see Phelps, [35], Section 3 for an exposition. Gossez considered this kind of multifunction in order to generalize to nonreflexive spaces some of the results previously known for reflexive spaces. In order to make this definition, we must introduce another concept due to Gossez: if S: $E \mapsto 2^{E^*}$, $\overline{S}$: $E^{**} \mapsto 2^{E^*}$ is defined by:

$$x^* \in \overline{S}x^{**} \quad \iff \quad \inf_{(s,s^*)\in G(S)} \langle s^* - x^*, \widehat{s} - x^{**}\rangle \geq 0.$$

Definition 25.1. S is *maximal monotone of type (D)* if S is maximal monotone and, for all $(x^{**}, x^*) \in G(\overline{S})$, there exists a bounded net (w_α, w_α^*) of elements of $G(S)$ such that $\widehat{w_\alpha} \to x^{**}$ in $w(E^{**}, E^*)$ and $\|w_\alpha^* - x^*\| \to 0$.

If E is reflexive then every maximal monotone multifunction is of type (D) and, even if E is not reflexive, subdifferentials are of type (D). Gossez proved this latter result in [26] by adapting one of Rockafellar's proof of the maximal monotonicity theorem. This proof involves some rather delicate functional analysis. We will prove a slight sharpening of Gossez's result in Theorem 35.3, using the formula for the biconjugate of a maximum that we develop in Theorem 33.3.

The multifunctions of type (FP) were introduced by Fitzpatrick–Phelps in [23], Section 3. Fitzpatrick–Phelps called these multifunctions "locally maximal monotone". We have renamed them since many people have found the original terminology confusing. The motivation for their introduction was as follows. If E is reflexive then every maximal monotone operator on E can be approximated by "nicer" maximal monotone operators using the Moreau-Yosida approximation. If E is nonreflexive then every subdifferential can also be approximated by "nicer" subdifferentials by using the operation of inf–convolution. So the question arises whether a general maximal monotone operators on a nonreflexive space can also be approximated by "nicer" maximal monotone operators in some appropriate sense. Fitzpatrick–Phelps defined an appropriate sense of approximation in [23], and showed that the multifunctions of type (FP) can be approximated by "nicer" maximal monotone operators in their sense.

Definition 25.2. A monotone multifunction S is *of type (FP)* provided the following holds: For any open convex subset U of E^* such that $U \cap R(S) \neq \emptyset$, if $(z, z^*) \in E \times U$ is such that

$$(w, w^*) \in G(S) \text{ and } w^* \in U \quad \Longrightarrow \quad \langle w - z, w^* - z^* \rangle \geq 0 \tag{25.2.1}$$

then $(z, z^*) \in G(S)$. (If we take $U := E^*$, we see that every multifunction of type (FP) is maximal monotone.)

Type (FP) multifunctions share the structural properties of type (D) multifunctions discussed above: if E is reflexive then every maximal monotone multifunction on E is of type (FP) (see [23], Proposition 3.3, p. 585). We will prove in Theorem 30.3 that, even if E is not reflexive, subdifferentials are of type (FP). Finally, it was proved in [24], Theorem 3.7, p. 67 that

$$S \text{ is maximal monotone and } R(S) = E^* \quad \Longrightarrow \quad S \text{ is of type (FP)}.$$

Remark 25.3. There is an example in Bauschke–Borwein, [6], Example 5.2 (originally due to Gossez) of a continuous linear skew operator that is not of type (FP). If we only ask that T be *positive* rather than *skew*, it is possible to give the much simpler example below, which is taken from Phelps–Simons, [36]. Let $E := \ell_1$, and T: $\ell^1 \mapsto E^* = \ell^\infty$ be defined by

$$(Tx)_n = \sum_{k \geq n} x_k \quad (x \in \ell_1).$$

Then T is positive (exercise!). Let $e := (1, 1, 1, \ldots) \in \ell^\infty$. Since c_0 is a closed subspace of ℓ_∞ and $e \notin c_0$, it follows from Theorem 4.4 that there exists $x^{**} \in E^{**}$ that vanishes on c_0 with $\langle e, x^{**} \rangle = 2$. Now let

$$U := \{x^* \in E^*\colon {x^*}_1 < \langle x^*, x^{**} \rangle\}.$$

U is a convex open subset of E^*. Suppose that $w \in E$. Then, by direct computation, $(Tw)_1 = \langle w, e \rangle$. Further, since $Tw \in c_0$, $\langle Tw, x^{**} \rangle = 0$. So if also $Tw \in U$ then $\langle w, e \rangle < 0$ and, consequently,

$$w \in E \text{ and } Tw \in U \quad \Longrightarrow \quad \langle w - 0, Tw - e \rangle = \langle w, Tw \rangle - \langle w, e \rangle > 0. \tag{25.3.1}$$

Now

$$\left[T(-1, 0, 0, \ldots)\right]_1 = (-1, 0, 0, \ldots)_1 = -1$$

and

$$\langle T(-1,0,0,\ldots), x^{**}\rangle = \langle(-1,0,0,\ldots), x^{**}\rangle = 0$$

Since $-1 < 0$, $T(-1,0,0,\ldots) \in U$, and so $U \cap R(T) \neq \emptyset$. Thus if T were of type (FP), it would follow from (25.3.1) that $(0,e) \in G(T)$, which is obviously impossible.

The multifunctions of type (FPV) were introduced by Fitzpatrick–Phelps in [24], p. 65 and Verona–Verona, [59], p. 268 by dualizing Definition 25.2. It is not known whether *every* maximal monotone multifunction is of type (FPV). See Theorem 26.1 for an explanation of why this could be a very hard problem. The significance of multifunctions of type (FPV) is, to some extent, explained by Section 26.

Definition 25.4. A monotone multifunction S is *of type (FPV)* provided the following holds: For any open convex subset U of E such that $U \cap D(S) \neq \emptyset$, if $(z,z^*) \in U \times E^*$ is such that

$$(w,w^*) \in G(S) \text{ and } w \in U \quad \Longrightarrow \quad \langle w-z, w^*-z^*\rangle \geq 0 \tag{25.4.1}$$

then $(z,z^*) \in G(S)$. (If we take $U := E$, we see that every multifunction of type (FPV) is maximal monotone.)

Type (FPV) multifunctions share the structural properties of type (D) and type (FP) multifunctions discussed above: if E is reflexive then every maximal monotone multifunction on E is of type (FPV) (see [23], Proposition 3.3, p. 585). We will prove in Theorem 31.3 that, even if E is not reflexive, subdifferentials are of type (FPV). In addition, we will prove in Theorem 38.2 that linear maximal monotone operators are of type (FPV). Finally, it was noted in [24], Theorem 3.10, p. 68 that

$$S \text{ is maximal monotone and } D(S) = E \quad \Longrightarrow \quad S \text{ is of type (FPV).}$$

We introduced multifunctions of type (NI) in [53], Definition 10, p. 183, motivated by some questions about the range of maximal monotone operators in nonreflexive spaces. Here is the definition, which should be compared with Lemma 8.1(c) — "NI" stands for "negative infimum".

Definition 25.5. Let $S\colon E \mapsto 2^{E^*}$ be maximal monotone. We say that S is *of type (NI)* if

$$(x^{**},x^*) \in E^{**} \times E^* \quad \Longrightarrow \quad \inf_{(s,s^*)\in G(S)} \langle s^*-x^*, \widehat{s}-x^{**}\rangle \leq 0. \tag{25.5.1}$$

We point out the following connection between multifunctions of type (NI) and multifunctions of type (D).

Lemma 25.6. *If S is maximal monotone of type (D) then S is maximal monotone of type (NI).*

Proof. Exercise!

Problem 25.7. If S is maximal monotone of type (NI) then does it necessarily follow that S is maximal monotone of type (D)? It was proved in Phelps–Simons, [36] that the answer to this question is in the affirmative if S is linear.

It follows from Lemma 25.6 that maximal monotone multifunctions of type (NI) share the structural properties of type (D), type (FP) and type (FPV) multifunctions discussed above: if E is reflexive then every maximal monotone multifunction on E is of type (NI) (see also Lemma 8.1(c)) and, even if E is not reflexive, subdifferentials are of type (NI).

Definition 25.8. We say that a multifunction $S\colon E \mapsto 2^{E^*}$ is *strongly maximal monotone* if S is monotone and whenever C is a nonempty $w(E, E^*)$–compact subset of E, $w^* \in E^*$ and

$$\text{for all } (y, y^*) \in G(S), \quad \text{there exists } w \in C \text{ such that} \quad \langle w - y, w^* - y^* \rangle \geq 0$$

then

$$\text{there exists } w \in C \text{ such that} \quad (w, w^*) \in G(S)$$

and, further, whenever C is a nonempty $w(E^*, E)$–compact subset of E^*, $w \in E$ and

$$\text{for all } (y, y^*) \in G(S), \quad \text{there exists } w^* \in C \text{ such that} \quad \langle w - y, w^* - y^* \rangle \geq 0$$

then

$$\text{there exists } w^* \in C \text{ such that} \quad (w, w^*) \in G(S).$$

Obviously, every strongly maximal monotone multifunction is maximal monotone. We will prove in Theorem 32.5 that subdifferentials are strongly maximal monotone, and in Theorem 38.5 that every maximal monotone (possibly discontinuous) positive linear operator is strongly maximal monotone.

These observations lead naturally to the following problem:

Problem 25.9. Is every maximal monotone multifunction strongly maximal monotone?

We conclude this section by mentioning a class of multifunctions that have a property of a more metric character. In the following definition, "ANA" stands for "almost negative alignment".

Definition 25.10. We say that $S\colon E \mapsto 2^{E^*}$ is *maximal monotone of type (ANA)* if, whenever $(x, x^*) \in E \times E^* \setminus G(S)$ then, for all $n \geq 1$, there exist $(w_n, w_n^*) \in G(S)$ such that $w_n \neq x$, $w_n^* \neq x^*$ and

$$\frac{\langle w_n - x, w_n^* - x^* \rangle}{\|w_n - x\| \|w_n^* - x^*\|} \to -1 \quad \text{as } n \to \infty.$$

It is clear from Corollary 10.4 that, if E is reflexive, then every maximal monotone multifunction on E is of type (ANA). We proved in [55] that subdifferentials are maximal monotone of type (ANA), and we will prove in Theorem 38.6 that continuous positive linear operators are maximal monotone of type (ANA). It is probably worth pointing out that if E is not reflexive then there is no hope of getting a result analogous to Corollary 10.4. From James's theorem, Theorem 4.11, there exists $w^* \in E^*$ that does not attain its norm on the unit ball of E. Define $T\colon E \mapsto E^*$ by $T := 0$. Then T is both a subdifferential and a continuous linear map, but there does not exist $(x, x^*) \in G(T)$ such that

$$x \neq w, \; x^* \neq w^* \quad \text{and} \quad \langle x - w, x^* - w^* \rangle = -\|x - w\| \|x^* - w^*\|. \quad (10.4.1)$$

Since $(x, x^*) \in G(T) \implies x^* = 0$, (10.4.1) would imply that $\langle x - w, w^* \rangle = \|x - w\| \|w^*\|$. Setting $b := (x - w)/\|x - w\|$, we would have $\|b\| = 1$ and $\langle b, w^* \rangle = \|w^*\|$, contradicting our choice of w^*.

These observations lead naturally to the following problem:

Problem 25.11. Is every maximal monotone multifunction of type (ANA)? (We do not even know what the situation is for *discontinuous* positive linear operators.)

26. The sum problem and the closure of the domain

The problem that has attracted the most interest since maximal monotonicity was introduced more than two decades ago and that has, so far, defied solution, is whether Rockafellar's original sum theorem is true for nonreflexive Banach spaces. Specifically, if E is not reflexive, $S\colon E \mapsto 2^{E^*}$ and $T\colon E \mapsto 2^{E^*}$ are maximal monotone and

$$D(S) \cap \operatorname{int} D(T) \neq \emptyset. \quad (26.0.1)$$

then is $S + T$ maximal monotone?

We have already observed in Section 25 that if $S\colon E \mapsto 2^{E^*}$ is maximal monotone and either E is reflexive or S is a subdifferential or S is linear then S is of type (FPV). We also made the comment that it could be a very hard problem to find an example of a maximal monotone multifunction that is not of type (FPV). Theorem 26.1 contains the explanation for that comment — its proof is borrowed from that of Fitzpatrick–Phelps, [23], Proposition 3.3, p. 585. A similar result was proved by Verona–Verona in [60].

Theorem 26.1. *Let $S\colon E \mapsto 2^{E^*}$ be maximal monotone and suppose that S has the property that if C is a nonempty closed convex subset of E, $T = N_C$ (see (8.1.1)) and (26.0.1) is satisfied then $S+T$ is maximal monotone. Then S is necessarily of type (FPV).*

Proof. Let U be an open convex subset of E such that $U \cap D(S) \neq \emptyset$, and $(z, z^*) \in U \times E^*$ satisfy (25.4.1). Fix $s \in U \cap D(S)$. Since the segment $[s, z]$ is compact, we can find $\epsilon > 0$ such that

$$C := [s, z] + \{x \in E\colon \; \|x\| \le \epsilon\} \subset U.$$

From (25.4.1),

$$(w, w^*) \in G(S) \text{ and } w \in C \quad \Longrightarrow \quad \langle w - z, w^* - z^* \rangle \ge 0. \tag{26.1.1}$$

Since $z \in C$,

$$(w, v^*) \in G(N_C) \quad \Longrightarrow \quad \langle w - z, v^* \rangle \ge 0. \tag{26.1.2}$$

Adding (26.1.1) and (26.1.2),

$$(w, w^*) \in G(S) \text{ and } (w, v^*) \in G(N_C) \quad \Longrightarrow \quad \langle w - z, w^* + v^* - z^* \rangle \ge 0,$$

that is to say,

$$(y, y^*) \in G(S + N_C) \quad \Longrightarrow \quad \langle y - z, y^* - z^* \rangle \ge 0. \tag{26.1.3}$$

Now $D(S) \cap \operatorname{int} D(N_C) = D(S) \cap \operatorname{int} C \ni s$ hence, by assumption, $S + N_C$ is maximal monotone. Thus, from (26.1.3), $(z, z^*) \in G(S + N_C)$, that is to say, $z^* \in Sz + N_C(z)$. Finally, since $z \in \operatorname{int} C$, $N_C(z) = \{0\}$, hence $z^* \in Sz$. This completes the proof that S is of type (FPV). ∎

It was pointed out in Problem 18.9 that it is unknown whether $\overline{D(S)}$ is necessarily convex when S is maximal monotone but E is not reflexive. Now it was proved in Theorem 18.6, that if E is reflexive and S is maximal monotone then $\overline{D(S)} = \overline{\operatorname{dom} \psi_S}$. (We recall that ψ_S was defined in Definition 15.1.) We shall prove in Theorem 26.3 that this result remains true even if E is not reflexive, provided that S is of type (FPV). So if it is a hard problem to find an example of a maximal monotone multifunction that is not of type (FPV), it is even harder to find one such that $\overline{D(S)} \neq \overline{\operatorname{dom} \psi_S}$. It is, course, then harder still to find one such that $\overline{D(S)}$ is not convex.

Our next result is valid for any nontrivial monotone multifunction.

Lemma 26.2. *Let $S\colon E \mapsto 2^{E^*}$ be monotone, $t \in D(S) \setminus \{0\}$, and $0 \in \operatorname{dom} \psi_S$. Let $\delta \in (0, 1/3)$, $V := \{x \in E\colon \; \|x\| < \delta\|t\|\}$ and $U := [0, t] + V$. Then* **either**

$$3V \cap D(S) \neq \emptyset$$

or *there exists $z^* \in E^*$ such that*

$$(w,w^*) \in G(S) \text{ and } w \in U \quad \Longrightarrow \quad \langle w, w^* - z^* \rangle \geq 0. \tag{26.2.1}$$

Proof. Let $M := 0 \vee \psi_S(0)$. Then $M \geq 0$ and

$$(w,w^*) \in G(S) \quad \Longrightarrow \quad \langle w, w^* \rangle + M(1 + \|w\|) \geq 0.$$

Since

$$w \in U \Longrightarrow \|w\| \leq 2\|t\|,$$

it follows that

$$(w,w^*) \in G(S) \text{ and } w \in U \quad \Longrightarrow \quad \langle w, w^* \rangle + M(1 + 2\|t\|) \geq 0. \tag{26.2.2}$$

Let

$$L := \frac{M(1 + 2\|t\|)}{\delta\|t\|}.$$

From the one–dimensional Hahn–Banach theorem, Corollary 1.2, there exists $z^* \in E^*$ such that

$$\|z^*\| = L \quad \text{and} \quad \langle -t, z^* \rangle = L\|t\|.$$

Suppose now that $3V \cap D(S) = \emptyset$. If $(w,w^*) \in G(S)$ and $w \in U$ then, since

$$\begin{aligned} U &= \big([0, 2\delta t] + V\big) \cup \big([2\delta t, t] + V\big) \\ &\subset 3V \cup \big([2\delta t, t] + V\big), \end{aligned}$$

it follows that there exists $\lambda \in [2\delta, 1]$ such that $\|w - \lambda t\| < \delta\|t\|$. Consequently,

$$\begin{aligned} -\langle w, z^* \rangle &= \lambda\langle -t, z^* \rangle - \langle w - \lambda t, z^* \rangle \\ &\geq 2\delta L\|t\| - L\delta\|t\| \\ &= L\delta\|t\| \\ &= M(1 + 2\|t\|). \end{aligned}$$

We now obtain (26.2.1) by combining this with (26.2.2). ▌

Theorem 26.3. *Let $S\colon E \mapsto 2^{E^*}$ be of type (FPV). Then*

$$\overline{D(S)} = \overline{\operatorname{co} D(S)} = \overline{\operatorname{dom} \chi_S} = \overline{\operatorname{dom} \psi_S}.$$

Proof. By virtue of the D–dom lemma, Lemma 15.2, it suffices to prove that $\operatorname{dom} \psi_S \subset \overline{D(S)}$. From Lemma 15.3, it suffices to prove that

$$0 \in \operatorname{dom} \psi_S \quad \Longrightarrow \quad 0 \in \overline{D(S)}.$$

This is obvious if $D(S) = \{0\}$, so we can suppose that there exists $t \in D(S) \setminus \{0\}$. Let $\delta \in (0, 1/3)$, and use the notation of Lemma 26.2. From Lemma 26.2, **either**

$$\text{there exists } x \in D(S) \text{ such that } \|x\| < 3\delta\|t\|$$

or

$$\text{there exist } z^* \in E^* \text{ satisfying (26.2.1).}$$

In the latter case, $(0, z^*) \in U \times E^*$ and, since $U \ni t$, $U \cap D(S) \neq \emptyset$. Since S is of type (FPV), it follows in this latter case that $(0, z^*) \in G(S)$, from which $0 \in D(S)$. If we now let $\delta \to 0$, we derive that $0 \in \overline{D(S)}$, as required. ∎

27. The closure of the range

In contrast to the situation explained in Section 26, we do know examples of maximal monotone multifunctions S such that $\overline{R(S)}$ is not convex. See, for instance, Fitzpatrick–Phelps, [24], Example 3.2, p. 63–64. It was essentially proved by Gossez in [26] (see Phelps, [35], Theorem 3.8, p. 22 for an exposition) that $\overline{R(S)}$ is convex if S is maximal monotone of type (D) (see Definition 25.1). It was proved in Fitzpatrick–Phelps, [23], Theorem 3.5, p. 585 that $\overline{R(S)}$ is also convex if S is of type (FP) (Definition 25.2). Finally, it was proved by Fitzpatrick–Phelps in [24] that $\overline{R(S)}$ is also convex if S is monotone and there exists $\eta > 0$ such that, for all $\lambda > 0$, the approximate resolvent $S + \lambda J_\eta$ is surjective. (We will explain the meaning of this in due course.) In this section, we shall show that, in all of the above situations, the statement "$\overline{R(S)}$ is convex" can be strengthened to "$\overline{R(S)} = \overline{\operatorname{dom} \xi_S}$". (We recall that ξ_S was defined in Definition 19.1.)

We first consider the type (FP) case. Here, the results are obtained by "dualizing" the appropriate results from Section 26. We start off with a result dual to Lemma 26.2.

Lemma 27.1. *Let $S\colon E \mapsto 2^{E^*}$ be monotone, $t^* \in R(S) \setminus \{0\}$ and $0 \in \operatorname{dom} \xi_S$. Let $\delta \in (0, 1/3)$, $V := \{x^* \in E^*\colon \|x^*\| < \delta\|t^*\|\}$ and $U := [0, t^*] + V$. Then* **either**

$$3V \cap R(S) \neq \emptyset$$

or *there exists $z \in E$ such that*

$$(w, w^*) \in G(S) \text{ and } w^* \in U \quad \Longrightarrow \quad \langle w - z, w^* \rangle \geq 0. \tag{27.1.1}$$

Proof. Let $M := 0 \vee \xi_S(0)$. Then $M \geq 0$ and

$$(w, w^*) \in G(S) \quad \Longrightarrow \quad \langle w, w^* \rangle + M(1 + \|w^*\|) \geq 0.$$

Since

$$w^* \in U \Longrightarrow \|w^*\| \leq 2\|t^*\|,$$

it follows that

$$(w, w^*) \in G(S) \text{ and } w^* \in U \quad \Longrightarrow \quad \langle w, w^* \rangle + M(1 + 2\|t^*\|) \geq 0. \tag{27.1.2}$$

Let

$$L := \frac{2M(1+2\|t^*\|)}{\delta\|t^*\|}.$$

From the definition of the norm of E^*, there exists $z \in E$ so that

$$\|z\| = L \quad \text{and} \quad \langle -z, t^*\rangle \geq 3L\|t^*\|/4.$$

Suppose now that $3V \cap R(S) = \emptyset$. If $(w, w^*) \in G(S)$ and $w^* \in U$ then, since

$$\begin{aligned} U &= \big([0, 2\delta t^*] + V\big) \cup \big([2\delta t^*, t^*] + V\big) \\ &\subset 3V \cup \big([2\delta t^*, t^*] + V\big), \end{aligned}$$

it follows that there exists $\lambda \in [2\delta, 1]$ such that $\|w^* - \lambda t^*\| < \delta\|t^*\|$. Consequently,

$$\begin{aligned} -\langle z, w^*\rangle &= \lambda\langle -z, t^*\rangle - \langle z, w^* - \lambda t^*\rangle \\ &\geq 2\delta(3L\|t^*\|/4) - L\delta\|t^*\| \\ &= L\delta\|t^*\|/2 \\ &= M(1+2\|t^*\|). \end{aligned}$$

We now obtain (27.1.1) by combining this with (27.1.2). ∎

Theorem 27.2. *Let* $S\colon E \mapsto 2^{E^*}$ *be of type (FP). Then*

$$\overline{R(S)} = \overline{\operatorname{dom} \xi_S}.$$

Proof. By virtue of (19.1.1), it suffices to prove that $\operatorname{dom} \xi_S \subset \overline{R(S)}$. Since we can show exactly as in Lemma 15.3 that *if* $T\colon E \mapsto 2^{E^*}$ *is nontrivial,* $w^* \in E^*$ *and* $S := T - w^*$ *then* $\operatorname{dom} \xi_S = \operatorname{dom} \xi_T - w^*$, it suffices to prove that

$$0 \in \operatorname{dom} \xi_S \quad \Longrightarrow \quad 0 \in \overline{R(S)}.$$

This is obvious if $R(S) = \{0\}$, so we can suppose that there exists $t^* \in R(S) \setminus \{0\}$. Let $\delta \in (0, 1/3)$, and use the notation of Lemma 27.1. From Lemma 27.1, **either**

$$\text{there exists } x^* \in R(S) \text{ such that } \|x^*\| < 3\delta\|t^*\|$$

or

$$\text{there exists } z \in E \text{ satisfying (27.1.1).}$$

In the latter case, $(z, 0) \in E \times U$ and, since $U \ni t^*$, $U \cap R(S) \neq \emptyset$. Since S is of type (FP), it follows in this latter case that $\underline{(z, 0)} \in G(S)$, from which $0 \in R(S)$. If we now let $\delta \to 0$, we derive that $0 \in \overline{R(S)}$, as required. ∎

Remark 27.3. The proofs of Theorems 26.3 and 27.2 were essentially obtained from an analysis of the proof of Fitzpatrick–Phelps, [23], Theorem 3.5. We point out another feature of Theorems 26.3 and 27.2: the full force of the definitions of *type (FPV)* and *type (FP)* are not used. It suffices for Theorem 26.3 to assume that: *for any open convex subset U of E such that $U \cap D(S) \neq \emptyset$, if $(z, z^*) \in U \times E^*$ is such that*

$$(w, w^*) \in G(S) \text{ and } w \in U \quad \Longrightarrow \quad \langle w - z, w^* - z^* \rangle \geq 0$$

then $z \in D(S)$. Similarly, it suffices for Theorem 27.2 to assume that: *for any open convex subset U of E^* such that $U \cap R(S) \neq \emptyset$, if $(z, z^*) \in E \times U$ is such that*

$$(w, w^*) \in G(S) \text{ and } w^* \in U \quad \Longrightarrow \quad \langle w - z, w^* - z^* \rangle \geq 0$$

then $z^* \in R(S)$.

We will need the following generalization of Lemma 10.1 to nonreflexive spaces. The proof follows exactly the same steps as that of Lemma 10.1, except that $(E \times E^*)^*$ can no longer be identified with $E \times E^*$.

Lemma 27.4. *Let $\emptyset \neq G \subset E \times E^*$. Then the conditions* (27.4.1) *and* (27.4.2) *are equivalent:*

$$\mu \in \mathcal{CO}(G) \quad \Longrightarrow \quad 2r(\mu) + \|p(\mu)\|^2 + \|q(\mu)\|^2 \geq 0. \tag{27.4.1}$$

$$\left.\begin{array}{l} \text{There exists } (x^*, x^{**}) \in E^* \times E^{**} \text{ such that} \\ (s, s^*) \in G \quad \Longrightarrow \\ \qquad 2\langle s^* - x^*, \widehat{s} - x^{**} \rangle \geq \|x^*\|^2 + \|x^{**}\|^2 + 2\langle x^*, x^{**} \rangle. \end{array}\right\} \tag{27.4.2}$$

We now deduce from Lemma 27.4 a simple proof of the main result about multifunctions of type (NI) (see [53], Theorem 12(a), p. 184). We shall use this result in Theorem 27.6.

Lemma 27.5. *Let $S\colon E \mapsto 2^{E^*}$ be maximal monotone of type (NI). Then there exists $(x^{**}, x^*) \in G(\overline{S})$ such that*

$$\|x^*\|^2 + \|x^{**}\|^2 + 2\langle x^*, x^{**} \rangle = 0. \tag{27.5.1}$$

Proof. It follows from Corollary 9.2(a) with $M := G(S)$ and Lemma 27.4 with $G := G(S)$ that there exists $(x^*, x^{**}) \in E^* \times E^{**}$ such that

$$\left.\begin{array}{l} (s, s^*) \in G(S) \quad \Longrightarrow \\ \qquad 2\langle s^* - x^*, \widehat{s} - x^{**} \rangle \geq \|x^*\|^2 + \|x^{**}\|^2 + 2\langle x^*, x^{**} \rangle. \end{array}\right\} \tag{27.5.2}$$

From the perfect square trick, Lemma 7.1,

$$(s, s^*) \in G(S) \quad \Longrightarrow \quad 2\langle s^* - x^*, \widehat{s} - x^{**} \rangle \geq 0,$$

hence $(x^{**}, x^*) \in G(\overline{S})$. Since S is of type (NI), (27.5.1) follows by taking the infimum over $(s, s^*) \in M$ in (27.5.2). ∎

It is worth noting that the maximality of S is not used in Lemma 27.5, only the monotonicity and the implication (25.5.1).

Theorem 27.6. *Let $S\colon E \mapsto 2^{E^*}$ be maximal monotone of type (D). Then*
$$\overline{R(S)} = \overline{\operatorname{dom}\xi_S}.$$

Proof. Arguing as in Theorem 27.2, it suffices to prove that
$$0 \in \operatorname{dom}\xi_S \quad\Longrightarrow\quad 0 \in \overline{R(S)}.$$
So let $0 \in \operatorname{dom}\xi_S$. Put $M := 0 \vee \xi_S(0)$. Then $M \geq 0$ and
$$(w, w^*) \in G(S) \quad\Longrightarrow\quad \langle w, w^*\rangle + M(1 + \|w^*\|) \geq 0. \tag{27.6.1}$$
Let $\varepsilon \in (0,1)$. We shall prove that
$$\text{there exists } w^* \in R(S) \text{ such that } \|w^*\| < \varepsilon, \tag{27.6.2}$$
which will give the required result. Choose $\lambda > 0$ so that
$$\lambda M < \frac{\varepsilon^2}{5} < 1, \tag{27.6.3}$$
and define $T\colon E \mapsto 2^{E^*}$ by $T := S/\lambda$. Then T is also maximal monotone of type (D) hence, from Lemma 25.6, maximal monotone of type (NI). From Lemma 27.5, there exists $(x^{**}, x^*) \in G(\overline{T})$ such that
$$\|x^*\|^2 + \|x^{**}\|^2 + 2\langle x^*, x^{**}\rangle = 0. \tag{27.6.4}$$
It is evident from the proof of Lemma 7.1 that $\|x^{**}\| = \|x^*\|$. Substituting this back in (27.6.4),
$$\langle x^*, x^{**}\rangle = -\|x^*\|^2. \tag{27.6.5}$$
Let $v^* := \lambda x^*$. Since $(x^{**}, x^*) \in G(\overline{T})$, $(x^{**}, v^*) \in G(\overline{S})$ and, using the fact that S is of type (D), there exists a bounded net (w_α, w_α^*) of elements of $G(S)$ such that
$$\widehat{w_\alpha} \to x^{**} \text{ in } w(E^{**}, E^*) \quad\text{and}\quad \|w_\alpha^* - v^*\| \to 0.$$
From (27.6.1), for all α,
$$\langle w_\alpha^*, \widehat{w_\alpha}\rangle + M(1 + \|w_\alpha^*\|) = \langle w_\alpha, w_\alpha^*\rangle + M(1 + \|w_\alpha^*\|) \geq 0$$
thus, passing to the limit,
$$\langle v^*, x^{**}\rangle + M(1 + \|v^*\|) \geq 0. \tag{27.6.6}$$
From (27.6.5),
$$\langle v^*, x^{**}\rangle = \lambda\langle x^*, x^{**}\rangle = -\lambda\|x^*\|^2 = -\|v^*\|^2/\lambda.$$
Substituting this back in (27.6.6),
$$\|v^*\|^2 - \lambda M\|v^*\| - \lambda M \leq 0$$
thus, arguing as in Lemma 18.5,
$$\|v^*\| < \varepsilon.$$
Since $\|w_\alpha^* - v^*\| \to 0$, there exists α such that $\|w_\alpha^*\| < \varepsilon$. Now $w_\alpha^* \in R(T)$, so this establishes (27.6.2), and completes the proof of Theorem 27.6. ▌

Theorem 27.6 suggests the following problem.

Problem 27.7. If S is maximal monotone of type (NI), is $\overline{R(S)}$ necessarily convex?

We now turn to the third situation in which it has been proved that $\overline{R(S)}$ is convex — when S is monotone and there exists $\eta > 0$ such that, for all $\lambda > 0$, the approximate resolvent $S + \lambda J_\eta$ is surjective. (See Fitzpatrick–Phelps, [24], Theorem 1.2, p. 54–56.) Now the statement "$S + \lambda J_\eta$ is surjective" means: for all $x^* \in E^*$, there exists $(w, w^*) \in G(S)$ such that

$$\|w\|^2 + \frac{2\langle w, w^* - x^*\rangle}{\lambda} + \frac{\|w^* - x^*\|^2}{\lambda^2} \leq 2\eta.$$

To avoid division by 0, let us suppose that $x^* \in E^* \setminus R(S)$. If now $0 < \lambda \leq 1/\eta$ then it follows from the above by dropping the $\|w\|^2$ term that

$$\frac{1}{2\lambda} \leq \frac{1 - \langle w, w^* - x^*\rangle}{\|w^* - x^*\|^2},$$

hence, letting $\lambda \to 0$,

$$\sup_{(w,w^*)\in G(S)} \frac{1 - \langle w, w^* - x^*\rangle}{\|w^* - x^*\|^2} = \infty.$$

Thus the result from [24] referred to above is extended by Theorem 27.8 below.

Theorem 27.8. *Let $S\colon E \mapsto 2^{E^*}$ be monotone and suppose that, for all $x^* \in E^* \setminus R(S)$, there exists $K \geq 0$ (depending on x^*) such that*

$$\sup_{(w,w^*)\in G(S)} \frac{K - \langle w, w^* - x^*\rangle}{\|w^* - x^*\|^2} = \infty.$$

Then

$$\overline{R(S)} = \overline{\operatorname{dom} \xi_S}.$$

Proof. By virtue of (19.1.1), it suffices to prove that $\operatorname{dom} \xi_S \subset \overline{R(S)}$. So let $x^* \in \operatorname{dom} \xi_S$. Put $M := 0 \vee \xi_S(x^*)$ and $N := M(1 + \|x^*\|) \geq 0$. Our aim is to prove that $x^* \in \overline{R(S)}$. This is obvious if $x^* \in R(S)$, so we can also suppose that $x^* \notin R(S)$ and define K as in the statement of the theorem. Let $\varepsilon \in (0, 1)$, and choose $\lambda > 0$ so that

$$\lambda(N + K) < \frac{\varepsilon^2}{5} < 1.$$

By hypothesis, there exists $(w, w^*) \in G(S)$ such that

$$\frac{K - \langle w, w^* - x^*\rangle}{\|w^* - x^*\|^2} \geq \frac{1}{\lambda},$$

from which

$$\|w^* - x^*\|^2 + \lambda\langle w, w^* - x^*\rangle - \lambda K \leq 0. \tag{27.8.1}$$

On the other hand,

$$\begin{aligned}\langle w, w^* - x^*\rangle + &N(1 + \|w^* - x^*\|) \\ &= \langle w, w^* - x^*\rangle + M(1 + \|x^*\|)(1 + \|w^* - x^*\|) \\ &\geq \langle w, w^* - x^*\rangle + M(1 + \|x^*\| + \|w^* - x^*\|) \\ &\geq \langle w, w^* - x^*\rangle + M(1 + \|w^*\|) \geq 0.\end{aligned}$$

Combining this with (27.8.1),

$$\|w^* - x^*\|^2 - \lambda N(1 + \|w^* - x^*\|) - \lambda K \leq 0,$$

i.e.,

$$\|w^* - x^*\|^2 - \lambda N\|w^* - x^*\| - \lambda(K + N) \leq 0.$$

Arguing exactly as in Theorem 27.6, $\|w^* - x^*\| < \varepsilon$. Letting $\varepsilon \to 0$, we see that $x^* \in \overline{R(S)}$, as required. ▮

Theorem 19.2 leads us to ask whether $\operatorname{int} R(S) = \operatorname{int}(\operatorname{dom} \xi_S)$ in the three situations that we have considered in this section. However, Borwein–Fitzpatrick–Vanderwerff have proved in [11], Theorem 3.1, p. 68 that if E is not reflexive then there exists a coercive, continuous convex function f on E such that $\operatorname{int} R(\partial f)$ is not convex.

We have already observed that subdifferentials are both maximal monotone of type (D) and type (FP). If S is a subdifferential then, for all $\eta,\ \lambda > 0$, $S + \lambda J_\eta$ is surjective. (This result, originally due to Gossez, follows from Theorem 35.3, Lemma 25.6 and Lemma 27.5.)

VII. Subdifferentials

28. The subdifferential of a sum

We start this chapter by developing in Theorem 28.2 the formula for the subdifferential of the sum of two convex functions, which we will use repeatedly in what follows. The main work for Theorem 28.2 is contained in Lemma 28.1, which we establish using the minimax technique. Actually, for the results of this chapter, we could equally well use the Attouch–Brézis theorem, Theorem 14.2, however we will need Lemma 28.1 in Theorem 37.1 in a situation where Theorem 14.2 cannot be used (since f is not lower semicontinuous).

Lemma 28.1. *Let $f\colon E \mapsto \mathrm{I\!R} \cup \{\infty\}$ and $g\colon E \mapsto \mathrm{I\!R} \cup \{\infty\}$ be convex, g be bounded above in some neighborhood of a point in* $\operatorname{dom} f$ *and*

$$f + g \geq 0 \quad \text{on } E. \tag{28.1.1}$$

Then there exists $y^* \in E^*$ such that

$$y \in \operatorname{dom} f \text{ and } z \in \operatorname{dom} g \quad \Longrightarrow \quad f(y) + g(z) + \langle z - y, y^* \rangle \geq 0. \tag{28.1.2}$$

Proof. By hypothesis, there exist $v \in \operatorname{dom} f$, $\eta > 0$ and $n \geq 1$ such that

$$\|w\| \leq \eta \quad \Longrightarrow \quad g(v + w) \leq n.$$

Let

$$M := \frac{f(v) + n}{\eta} \geq \frac{f(v) + g(v)}{\eta} \geq 0.$$

We now prove that

$$(y, z) \in \operatorname{dom} f \times \operatorname{dom} g \quad \Longrightarrow \quad f(y) + g(z) + M\|z - y\| \geq 0. \tag{28.1.3}$$

So let $(y, z) \in \operatorname{dom} f \times \operatorname{dom} g$. If $z = y$ then (28.1.3) is immediate from (28.1.1), so we can and will assume that $z \neq y$. Let

$$\lambda := \frac{\eta}{\|z - y\|} > 0 \quad \text{and} \quad w := \lambda(y - z).$$

Clearly

$$\frac{v+\lambda y}{1+\lambda} = \frac{v+(w+\lambda z)}{1+\lambda} = \frac{(v+w)+\lambda z}{1+\lambda}.$$

Thus, from (28.1.1),

$$f\left(\frac{v+\lambda y}{1+\lambda}\right) + g\left(\frac{(v+w)+\lambda z}{1+\lambda}\right) \geq 0$$

and so, using the convexity of f and g,

$$f(v) + \lambda f(y) + g(v+w) + \lambda g(z) \geq 0.$$

Since $\|w\| = \eta$, $g(v+w) \leq n$. (28.1.3) now follows by substituting in the values of $\lambda = \eta/\|z-y\|$ and $M = (f(v)+n)/\eta$. Let $A := \operatorname{dom} f \times \operatorname{dom} g$, and

$$B := \{y^* \in E^*\colon \ \|y^*\| \leq M\},$$

with the topology $w(E^*, E)$. From the Banach–Alaoglu theorem, Theorem 4.1, B is compact. Define $h\colon\ A \times B \mapsto \mathbb{R}$ by

$$h((y,z),y^*) := f(y) + g(z) + \langle z-y, y^*\rangle.$$

It follows from (28.1.3) and the one–dimensional Hahn–Banach theorem, Corollary 1.2, that

$$\inf_A \max_B h \geq 0.$$

The function h is convex on A, and affine and continuous on B. Thus, from the minimax theorem, Theorem 3.1,

$$\max_B \inf_A h \geq 0,$$

that is to say, there exists $y^* \in B$ satisfying (28.1.2). ▌

When we defined the concept of the *subdifferential* of a convex function f in Section 8, we assumed that f was lower semicontinuous. For this section, and this section only, we relax that hypothesis. Theorem 28.2 clearly implies the result in Phelps, [34], Theorem 3.16, p. 47, where it is assumed that e is continuous at a point in $\operatorname{dom} e \cap \operatorname{dom} f$. Further, if e is lower semicontinuous and $\operatorname{dom} f \cap \operatorname{int} \operatorname{dom} e \neq \emptyset$ then the boundedness hypothesis of Theorem 28.2 follows automatically from the dom lemma, Lemma 12.2.

Theorem 28.2. *Let $e\colon E \mapsto \mathbb{R} \cup \{\infty\}$ and $f\colon E \mapsto \mathbb{R} \cup \{\infty\}$ be convex, and e be bounded above in some neighborhood of a point in $\operatorname{dom} f$. Then $\partial(e+f) = \partial e + \partial f$.*

Proof. We leave as an exercise the proof of the inclusion

$$G(\partial(e+f)) \supset G(\partial e + \partial f).$$

To prove the opposite inclusion, suppose that $(x, x^*) \in G(\partial(e+f))$. So $x \in \operatorname{dom}(e+f)$ and

$$u \in E \quad \Longrightarrow \quad e(u) + f(u) + \langle x - u, x^* \rangle - e(x) - f(x) \geq 0. \tag{28.2.1}$$

Define $g\colon E \mapsto \mathbb{R} \cup \{\infty\}$ by

$$g(z) := e(z) + \langle x - z, x^* \rangle - e(x) - f(x) \quad (z \in E).$$

Then (28.1.1) follows from (28.2.1). From Lemma 28.1, there exists $y^* \in E$ satisfying (28.1.2). Plugging in the definition of g, for all $z \in \operatorname{dom} e$ and $y \in \operatorname{dom} f$,

$$e(z) + \langle x - z, x^* - y^* \rangle - e(x) + f(y) + \langle x - y, y^* \rangle - f(x) \geq 0.$$

Putting $z = x$ in this gives

$$y^* \in \partial f(x),$$

and putting $y = x$ gives

$$x^* - y^* \in \partial e(x).$$

Consequently, $x^* = (x^* - y^*) + y^* \in (\partial e + \partial f)(x)$, and so $(x, x^*) \in G(\partial e + \partial f)$. This completes the proof of Theorem 28.2. ∎

29. Subdifferentials are maximal monotone

For the remainder of this chapter, we consider generalizations of Rockafellar's maximal monotonicity theorem (first proved in [44]):

$$f \in \mathcal{PCLSC}(E) \quad \Longrightarrow \quad \partial f\colon E \mapsto 2^{E^*} \text{ is maximal monotone.}$$

In order to establish this, we must prove that if

$$(z, z^*) \in E \times E^* \setminus G(\partial f)$$

then there exists $(w, w^*) \in G(\partial f)$ such that

$$\langle w - z, w^* - z^* \rangle < 0.$$

This result can easily be deduced from Corollary 29.5 using the substitution

$$g(x) := f(x + z) - \langle x, z^* \rangle \quad (x \in E).$$

It turns out to be quite difficult to produce $x^* \in E^*$ satisfying the *strict* inequality in (29.5.1). Paradoxically, it is easier to prove Theorem 29.4, which is a much stronger result. Theorem 29.4 is stated in such a way as to lead easily to the result in the next section that subdifferentials are maximal monotone of type (FP). In order to obtain Corollary 29.5, it suffices to take $S := \| \ \|$ and $m = M = 1$ in Theorem 29.4. Theorem 29.4 appears in our paper [49], Lemma 2, p. 468–469, with a proof based on the properties of directional derivatives and sublinear functionals. In accordance with the general philosophy of these notes, we give here a proof using the formula for the subdifferential of a sum, which we established in Theorem 28.2 using the minimax technique, and Ekeland's variational principle.

Ekeland's variational principle is a result on complete metric spaces which has had a large number of applications to nonlinear analysis. We refer the reader to Ekeland's survey article, [20], for a description of some of these. The Banach space case, as in the statement of Theorem 29.1 below, is proved in Phelps, [34], Lemma 3.13, p. 45. Note that it does *not* require f to be convex.

Theorem 29.1. *Let $f\colon E \mapsto \mathbb{R} \cup \{\infty\}$ be a lower semicontinuous function $\alpha, \beta > 0$, $y \in \operatorname{dom} f$ and $f(y) \le \inf_E f + \alpha\beta$. Then there exists $x \in E$ such that $f(x) \le f(y)$ (hence $x \in \operatorname{dom} f$), $\|x - y\| \le \alpha$ and*

$$z \in \operatorname{dom} f \quad \Longrightarrow \quad f(z) + \beta\|z - x\| - f(x) \ge 0.$$

We now use the minimax technique to deduce the Brøndsted–Rockafellar theorem (see [14], p. 608 and Phelps, [34], Theorem 3.17, p. 48) from Theorem 29.1:

Corollary 29.2. *Let $f \in \mathcal{PCLSC}(E)$, $\alpha, \beta > 0$, $y \in \operatorname{dom} f$ and $f(y) \le \inf_E f + \alpha\beta$. Then there exists $(x, z^*) \in G(\partial f)$ such that $\|x - y\| \le \alpha$, $f(x) \le f(y)$ and $\|z^*\| \le \beta$.*

Proof. Let x be as in Theorem 29.1. Let $A := \operatorname{dom} f$, and

$$B := \{z^* \in E^*\colon \|z^*\| \le \beta\}$$

with the topology $w(E^*, E)$. From the Banach–Alaoglu theorem, Theorem 4.1, B is compact. Define $h\colon A \times B \mapsto \mathbb{R}$ by

$$h(z, z^*) := f(z) + \langle x - z, z^* \rangle - f(x).$$

Then, from Theorem 29.1 and the one–dimensional Hahn–Banach theorem, Corollary 1.2,

$$\inf_A \max_B h \ge 0.$$

The function h is convex on A, and affine and continuous on B. Thus from the minimax theorem, Theorem 3.1,

$$\max_B \inf_A h \ge 0,$$

and so there exists $z^* \in B$ such that

$$z \in \operatorname{dom} f \quad \Longrightarrow \quad f(z) + \langle x - z, z^* \rangle - f(x) \ge 0. \ \blacksquare$$

We leave the proof of the following result as an exercise.

Lemma 29.3. *Let $T : E \mapsto \mathbb{R}$ be continuous and sublinear, $x \in E$ and $y^* \in E^*$. Then $y^* \in \partial T(x)$ if, and only if, $y^* \leq T$ on E and $\langle x, y^* \rangle = T(x)$.*

We now come to the main result of this section, Theorem 29.4. In the case where the sublinear functional S introduced therein is identical with $\| \ \|$, we can give simple graphical interpretations of the constant K introduced in (29.4.4). In this case,

$$K := \sup_{y \in F} \frac{\lambda - g(y)}{\|y\|}.$$

So K is the supremum of the slopes of the line segments from $(0, \lambda)$ to points on the graph of g that lie below λ.

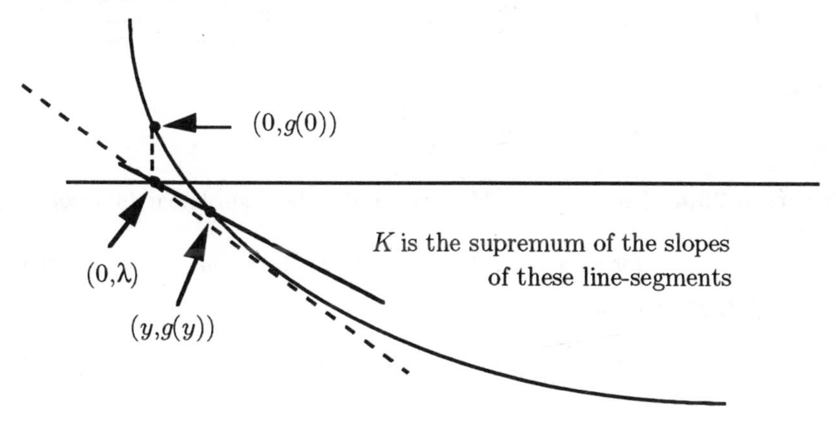

We can also think of K as the infimum of the slopes of the downward cones with vertex $(0, \lambda)$ that miss the graph of g.

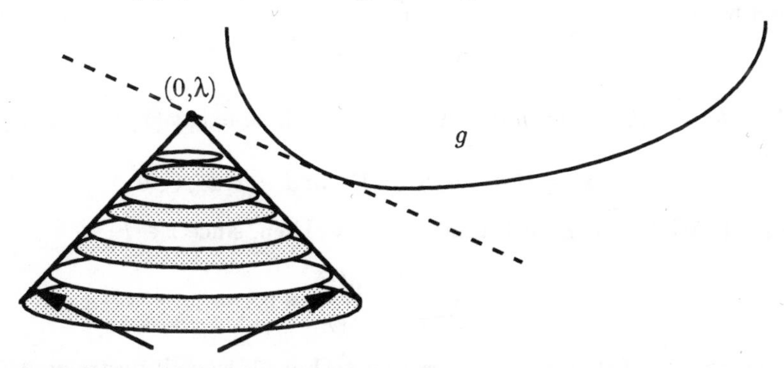

K is the infimum of the slopes of these cones

The constant K also has a less obvious interpretation, namely as the infimum of the slopes of the subtangents to the graph of g that dominate $(0, 0)$. This follows from Simons, [50], Theorem 5.4, p. 1384. In the case where S is a general sublinear functional satisfying (29.4.1), K is much harder to visualize.

A certain amount of effort in Theorem 29.4 below is devoted to showing that $K < \infty$. We have already observed in Remark 6.3 that the fact that $K < \infty$ leads to the formula ${}^*g^* = g$. We point out here that if we assume this formula (as many readers will be happy to do) then there is a much shorter proof that $K < \infty$. To see this, we observe from (29.4.2) that we then have

$$\lambda < g(0) = {}^*g^*(0) = -\inf_{E^*} g^*.$$

Consequently, there exists $x^* \in E^*$ such that $g^*(x^*) \le -\lambda$. Then, from (29.4.1),

$$\text{for all } y \in E, \quad \frac{\lambda - g(y)}{S(-y)} \le \frac{\langle -y, x^*\rangle}{S(-y)} \le \|x^*\| \frac{\|-y\|}{S(-y)} \le \frac{\|x^*\|}{m},$$

and so

$$K \le \frac{\|x^*\|}{m} < \infty.$$

Theorem 29.4. *Let $0 < m \le M < \infty$, and S be a sublinear functional on E such that*

$$\text{for all } x \in E, \quad m\|x\| \le S(x) \le M\|x\|. \tag{29.4.1}$$

Let $g \in \mathcal{PCLSC}(E)$ and

$$\inf_E g < \lambda < g(0)\ (\le \infty). \tag{29.4.2}$$

Let

$$F := \{y \in E\colon\ g(y) \le \lambda\} \ne \emptyset, \tag{29.4.3}$$

and write

$$K := \sup_{y \in F} \frac{\lambda - g(y)}{S(-y)}. \tag{29.4.4}$$

Also, let $\varepsilon \in (0,1)$. Then $0 < K < \infty$, and there exists $(x, x^*) \in G(\partial g)$ such that

$$x^* \le (1+\varepsilon)KS \text{ on } E \quad \text{and} \quad \langle x, x^*\rangle < 0. \tag{29.4.5}$$

Proof. We fix $z \in E$ such that $g(z) < \lambda$. Then, since $z \in F$,

$$K \ge \frac{\lambda - g(z)}{S(-z)} > 0.$$

Since the proof that $K < \infty$ is rather technical, we will postpone it for the moment, and continue with the rest of the proof. Clearly

$$y \in F \quad \Longrightarrow \quad \frac{\lambda - g(y)}{S(-y)} \le K \quad \Longrightarrow \quad \lambda \le g(y) + KS(-y)$$

and

$$y \in E \setminus F \quad \Longrightarrow \quad g(y) \ge \lambda \quad \Longrightarrow \quad \lambda \le g(y) + KS(-y).$$

Combining together these two inequalities:

$$\lambda \leq \inf_E (g + KS^-), \tag{29.4.6}$$

where the sublinear functional S^- is defined by $S^-(x) := S(-x) \quad (x \in E)$. Since $(1 - \varepsilon m/2M)K < K$, from (29.4.4), there exists $y \in F$ (from which $\|y\| > 0$) such that

$$\frac{\lambda - g(y)}{S(-y)} > \left(1 - \frac{\varepsilon m}{2M}\right) K.$$

Clearing of fractions and using (29.4.1) and (29.4.6),

$$g(y) + KS^-(y) \leq \lambda + \frac{\varepsilon m K S^-(y)}{2M} \leq \inf_E (g + KS^-) + \frac{\|y\|}{2} \varepsilon m K.$$

Noting from (29.4.1) that S^- is continuous, we derive from the Brøndsted–Rockafellar theorem, Corollary 29.2, with $f := g + KS^-$, $\alpha := \|y\|/2$ and $\beta := \varepsilon m K$, that there exists $(x, z^*) \in G(\partial(g + KS^-))$ such that

$$\|x - y\| \leq \frac{\|y\|}{2} \quad \text{and} \quad \|z^*\| \leq \varepsilon m K. \tag{29.4.7}$$

We first observe from (29.4.7) that $x \neq 0$. From the formula for the subdifferential of a sum, Theorem 28.2, there exist $x^* \in \partial g(x)$ and $y^* \in \partial(KS^-)(x)$ such that $x^* + y^* = z^*$. From Lemma 29.3,

$$y^* \leq KS^- \text{ on } E \quad \text{and} \quad \langle x, y^* \rangle = KS(-x)$$

hence

$$x^* = -y^* + z^* \leq KS + \varepsilon m K \| \ \| \leq (1 + \varepsilon) KS \text{ on } E$$

and

$$\langle x, x^* \rangle = -\langle x, y^* \rangle + \langle x, z^* \rangle \leq -KS(-x) + \varepsilon m K \|x\| \leq -(1 - \varepsilon) KS(-x).$$

Since $x \neq 0$, we obtain finally from (29.4.1) that $S(-x) > 0$, hence $\langle x, x^* \rangle < 0$. This completes the proof of (29.4.5). We now fill in the details of the proof that $K < \infty$. Since g is lower semicontinuous, we can choose θ, $\eta > 0$ such that

$$g(u) \leq g(z) - 1 \quad \Longrightarrow \quad S(z - u) \geq \theta \tag{29.4.8}$$

and

$$y \in F \quad \Longrightarrow \quad S(-y) \geq \eta. \tag{29.4.9}$$

Let y be an arbitrary element of F. If $g(y) > g(z) - 1$ then, from (29.4.9),

$$\frac{\lambda - g(y)}{S(-y)} \leq \frac{\lambda - g(z) + 1}{\eta} < \infty.$$

If, on the other hand, $g(y) \leq g(z) - 1$ let

$$\gamma := \frac{1}{g(z) - g(y)} \in (0, 1].$$

Then

$$g(\gamma y + (1-\gamma)z) \le \gamma g(y) + (1-\gamma)g(z) = g(z) - 1$$

and so, using (29.4.8) with $u := \gamma y + (1-\gamma)z$,

$$\gamma S(z-y) = S(z - \gamma y - (1-\gamma)z) \ge \theta$$

from which

$$g(z) - g(y) = \frac{1}{\gamma} \le \frac{S(z-y)}{\theta}.$$

Thus

$$\lambda - g(y) \le \lambda - g(z) + \frac{S(z-y)}{\theta} \le \lambda - g(z) + \frac{S(z) + S(-y)}{\theta}.$$

Consequently,

$$\frac{\lambda - g(y)}{S(-y)} \le \frac{\lambda - g(z)}{S(-y)} + \frac{S(z)}{\theta S(-y)} + \frac{1}{\theta}$$

and, using (29.4.9) again,

$$\frac{\lambda - g(y)}{S(-y)} \le \frac{\lambda - g(z)}{\eta} + \frac{S(z)}{\theta\eta} + \frac{1}{\theta} < \infty.$$

Thus $K < \infty$, as required. This completes the proof of Theorem 29.4. ▌

For many (but not all) purposes, the following consequence of Theorem 29.4 is adequate.

Corollary 29.5. *Let $g \in \mathcal{PCLSC}(E)$ and $\inf_E g < g(0)$ $(\le \infty)$. Then there exists $(x, x^*) \in G(\partial g)$ such that*

$$\langle x, x^* \rangle < 0. \tag{29.5.1}$$

Proof. We can find λ such that $\inf_E g < \lambda < g(0)$, apply Theorem 29.4 with $S := \|\ \|$ and $\varepsilon = 1/2$, and then throw away all parts of the conclusion that involve λ or K. ▌

30. Subdifferentials are of type (FP)

We shall prove in Theorem 30.3 the result already advertised in Section 25 that *subdifferentials are of type (FP)*. The material in this section is a simplification of the analysis in our paper [49].

Lemma 30.1. *Let $g \in \mathcal{PCLSC}(E)$ and $\inf_E g < g(0)$ $(\leq \infty)$. Let C be a $w(E^*, E)$–compact convex subset of E^*, $0 \in \operatorname{int} C$ and $R(\partial g) \cap \operatorname{int} C \neq \emptyset$. Then there exists $(x, x^*) \in G(\partial g)$ such that*

$$x^* \in C \quad \text{and} \quad \langle x, x^* \rangle < 0. \tag{30.1.1}$$

Proof. Define the sublinear functional S on E by

$$S(x) := \max\langle x, C\rangle \quad (x \in E).$$

S clearly satisfies (29.4.1). Pick $v^* \in R(\partial g) \cap \operatorname{int} C$. Since $v^* \in \operatorname{int} C$, there exists $\varepsilon \in (0,1)$ such that $(1+\varepsilon)v^* \in C$. Write $N := 1/(1+\varepsilon)$. We now show that we can choose λ satisfying (29.4.2) such that, if F is as in (29.4.3) then

$$\inf_F (g + NS^-) \geq \lambda. \tag{30.1.2}$$

If $\inf_E g = -\infty$, we choose $v \in (\partial g)^{-1}v^*$ and then λ so that $-\infty < \lambda < g(0)$ and $\lambda \leq g(v) - \langle v, v^*\rangle$. Then, for all $y \in E$,

$$\begin{aligned} g(y) + NS(-y) &\geq g(y) + N\langle -y, (1+\varepsilon)v^*\rangle \\ &\geq g(v) + \langle y - v, v^*\rangle - \langle y, v^*\rangle \\ &= g(v) - \langle v, v^*\rangle \geq \lambda, \end{aligned}$$

which gives (30.1.2). If, on the other hand, $\inf_E g > -\infty$, we first choose λ_0 so that

$$\inf_E g < \lambda_0 < g(0).$$

Since g is lower semicontinuous, we can then choose $\delta > 0$ so that

$$g(y) \leq \lambda_0 \quad \Longrightarrow \quad \|y\| \geq \delta. \tag{30.1.3}$$

Finally, we can choose λ so that

$$\inf_E g < \lambda \leq \lambda_0 \quad \text{and} \quad \lambda \leq \inf_E g + Nm\delta,$$

where m satisfies (29.4.1). From (30.1.3) and (29.4.1),

$$y \in F \quad \Longrightarrow \quad \|y\| \geq \delta \quad \Longrightarrow \quad S(-y) \geq m\delta,$$

from which

$$\inf_F (g + NS^-) \geq \inf_F g + \inf_F NS^- \geq \inf_E g + Nm\delta \geq \lambda,$$

which gives (30.1.2) again. We now define K as in (29.4.4). It is immediate from (30.1.2) that $K \leq N$. From Theorem 29.4, there exists $(x, x^*) \in G(\partial g)$ such that

$$x^* \leq (1+\varepsilon)KS \text{ on E} \quad \text{and} \quad \langle x, x^*\rangle < 0.$$

Since $K \leq N$ and $N = 1/(1+\varepsilon)$, we have $(1+\varepsilon)K \leq 1$. Consequently,

$$x^* \leq S \text{ on } E.$$

We now obtain from Theorem 4.8 that $x^* \in C$. This completes the proof of (30.1.1), and also that of Lemma 30.1. ▌

Corollary 30.2. *Let $g \in \mathcal{PCLSC}(E)$ and $\inf_E g < g(0)$ $(\leq \infty)$. Let V be a convex open subset of E^* such that*

$$V \ni 0 \quad \text{and} \quad V \cap R(\partial g) \neq \emptyset.$$

Then there exists $(x, x^*) \in G(\partial g)$ such that

$$x^* \in V \quad \text{and} \quad \langle x, x^* \rangle < 0.$$

Proof. Fix $v^* \in V \cap R(\partial g)$. The result now follows from Lemma 30.1 with

$$C := [0, v^*] + \{x^* \in E^*\colon\ \|x^*\| \leq \delta\},$$

and $\delta > 0$ so small that $C \subset V$. ▌

Theorem 30.3. *Let $f \in \mathcal{PCLSC}(E)$. Then $\partial f\colon\ E \mapsto 2^{E^*}$ is of type (FP).*

Proof. Putting Definition 25.2 in contrapositive form, what we must prove is that if U is a convex open subset of E^* such that

$$U \cap R(\partial f) \neq \emptyset$$

and

$$(z, z^*) \in E \times U \setminus G(\partial f)$$

then there exists $(w, w^*) \in G(\partial f)$ such that

$$w^* \in U \quad \text{and} \quad \langle w - z, w^* - z^* \rangle < 0.$$

This can be deduced easily from Corollary 30.2 using the substitutions

$$V := U - z^*$$

and

$$g(x) := f(x + z) - \langle x, z^* \rangle \quad (x \in E). \text{ ▌}$$

31. Subdifferentials are of type (FPV)

We shall prove in Theorem 31.3 the result already advertised in Section 25 that *subdifferentials are of type (FPV)*. This was first proved by Fitzpatrick–Phelps in [24], Corollary 3.4, p. 66 and Verona–Verona in [59], Theorem 3, p. 269.

We point out that Theorem 29.4 does not give any specific information about the location of x other than that $x \neq 0$. In Lemma 31.1, we give a result in which we do have more specific information about this. The price we pay is that we have to sacrifice the estimate for $\|x^*\|$ that we had in (29.4.5). Lemma 31.1 was suggested by the result of Verona–Verona referred to above.

Lemma 31.1. *Let $k \in \mathcal{PCLSC}(E)$, $v \in E$ and $k(v) < k(0)$ $(\le \infty)$. Let C be a closed convex set such that*

$$0 \in C \quad \text{and} \quad v \in \operatorname{int} C.$$

Then there exists $(x, z^*) \in G(\partial k)$ such that

$$x \in C \quad \text{and} \quad \langle x, z^* \rangle < 0.$$

Proof. Since $v \in C$ and $0 \in C$, with I_C defined as in Section 8,

$$\inf_E (k + I_C) \le (k + I_C)(v) = k(v) < k(0) = (k + I_C)(0).$$

We apply Corollary 29.5 to $g := k + I_C$ and obtain $(x, x^*) \in G(\partial(k + I_C))$ such that $\langle x, x^* \rangle < 0$. From the formula for the subdifferential of a sum, Theorem 28.2, there exist $y^* \in \partial I_C(x) = N_C(x)$ and $z^* \in \partial k(x)$ such that $x^* = y^* + z^*$. Since $y^* \in N_C(x)$ and $0 \in C$,

$$x \in C \quad \text{and} \quad \langle x, y^* \rangle \ge \langle 0, y^* \rangle = 0.$$

Consequently,

$$\langle x, z^* \rangle = \langle x, x^* \rangle - \langle x, y^* \rangle < 0.$$

This completes the proof of Lemma 31.1. ▌

Corollary 31.2. *Let $k \in \mathcal{PCLSC}(E)$ and $\inf_E k < k(0)$ $(\le \infty)$. Let V be a convex open subset of E such that*

$$V \ni 0 \quad \text{and} \quad V \cap \operatorname{dom} k \ne \emptyset.$$

Then there exists $(x, x^*) \in G(\partial k)$ such that

$$x \in V \quad \text{and} \quad \langle x, x^* \rangle < 0.$$

Proof. We first show that

$$\text{there exists } v \in V \text{ such that } k(v) < k(0). \tag{31.2.1}$$

Fix $y \in V \cap \operatorname{dom} k$ and $z \in E$ such that $k(z) < k(0)$. If $k(0) = \infty$ then $k(y) < k(0)$, and we write $v := y$. If, on the other hand, $k(0) \in \mathbb{R}$ then, for all $\lambda \in (0, 1]$, $k(\lambda z) < k(0)$. We choose $\lambda \in (0, 1]$ so small that $\lambda z \in V$, and write $v := \lambda z$. In either case, we have established (31.2.1). The result now follows from Lemma 31.1 with

$$C := [0, v] + \{x \in E \colon \|x\| \le \delta\},$$

and $\delta > 0$ so small that $C \subset V$. ▌

Theorem 31.3. *Let $f \in \mathcal{PCLSC}(E)$. Then $\partial f\colon E \mapsto 2^{E^*}$ is of type (FPV).*

Proof. Putting Definition 25.4 in contrapositive form, what we must prove is that if U is a convex open subset of E such that

$$U \cap D(\partial f) \neq \emptyset$$

and

$$(z, z^*) \in U \times E^* \setminus G(\partial f)$$

then there exists $(w, w^*) \in G(\partial f)$ such that

$$w \in U \quad \text{and} \quad \langle w - z, w^* - z^* \rangle < 0.$$

This can be deduced easily from Corollary 31.2 using the substitutions

$$V := U - z$$

and

$$k(x) := f(x + z) - \langle x, z^* \rangle \quad (x \in E). \blacksquare$$

Remark 31.4. We do not use the full force of the hypothesis "$v^* \in R(\partial g)$" in the proof of Lemma 30.1. All we need, in fact, is that,

$$\inf_E (g - v^*) > -\infty,$$

or equivalently

$$\sup_E (v^* - g) < \infty.$$

This can be rewritten $g^*(v^*) < \infty$, that is to say $v^* \in \operatorname{dom} g^*$. So the hypothesis

$$R(\partial g) \cap \operatorname{int} C \neq \emptyset$$

in the statement of Lemma 30.1 can be replaced by

$$\operatorname{dom} g^* \cap \operatorname{int} C \neq \emptyset.$$

This observation makes Lemma 31.1 look more like a "dual version" of Lemma 30.1. We refer the reader to our paper [55] for a more extensive discussion of "dual results".

32. Subdifferentials are strongly maximal monotone

The main result in this section is Theorem 32.5, in which we show, as advertised in Section 25, that subdifferentials are strongly maximal monotone. This was first proved in our paper [50], Theorems 6.1 and 6.2, p. 1386, using the properties of sublinear functionals and directional derivatives. The proof that we give here depends on the formula for the subdifferential of a sum, which we established in Theorem 28.2 using the minimax technique.

Our first result is a generalization of Corollary 29.5, in which $\{0\}$ is replaced by any nonempty $w(E, E^*)$–compact convex subset of E. Since the constant K defined in (32.1.3) can be obtained from the equivalent formula

$$K = \sup_{y \in F,\ w \in C} \frac{\lambda - g(y)}{\|y - w\|},$$

we can represent it graphically by the following picture.

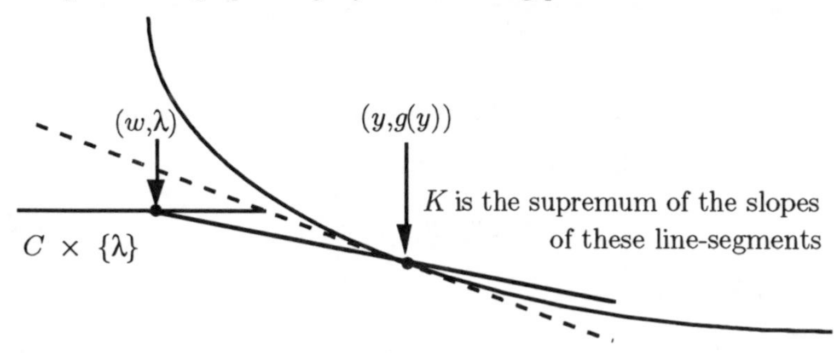

Lemma 32.1. *Let C be a nonempty $w(E, E^*)$–compact convex subset of E, $g \in \mathcal{PCLSC}(E)$ and, for all $w \in C$, $0 \notin \partial g(w)$. Then there exists $(x, x^*) \in G(\partial g)$ such that*

$$\text{for all } w \in C, \quad \langle x - w, x^* \rangle < 0. \tag{32.1.1}$$

Proof. From Corollary 4.5, the function g is $w(E, E^*)$–lower semicontinuous, hence there exists $w \in C$ such that

$$g(w) = \min_C g \ (\le \infty).$$

By hypothesis, $0 \notin \partial g(w)$, from which $\inf_E g < g(w)$. Combining this with the equality above, $\inf_E g < \min_C g$. We now choose λ so that

$$\inf_E g < \lambda < \min_C g$$

and let

$$F := \{y \in E\colon\ g(y) \le \lambda\}.$$

Then F is nonempty, convex and closed and $F \cap C = \emptyset$. Thus, from Corollary 4.6,

$$\eta := \inf \|F - C\| > 0.$$

Define d_C: $E \mapsto \mathbb{R}$ by $d_C(y) := \inf \|y - C\|$. We note then that

$$y \in F \implies d_C(y) \geq \eta > 0. \tag{32.1.2}$$

Let

$$K := \sup_{y \in F} \frac{\lambda - g(y)}{d_C(y)}. \tag{32.1.3}$$

Clearly $K > 0$. In fact

$$K < \infty. \tag{32.1.4}$$

Since the proof of (32.1.4) is rather technical, we will postpone it for the moment, and continue with the rest of the proof. Clearly,

$$y \in F \implies \frac{\lambda - g(y)}{d_C(y)} \leq K \implies \lambda \leq g(y) + K d_C(y)$$

and

$$y \in E \setminus F \implies g(y) \geq \lambda \implies \lambda \leq g(y) + K d_C(y).$$

Combining together these two inequalities:

$$\lambda \leq \inf_E (g + K d_C). \tag{32.1.5}$$

Now choose δ so that

$$0 < \delta < \frac{1}{2} \quad \text{and} \quad \delta \operatorname{diam} C < \frac{\eta}{4}. \tag{32.1.6}$$

Since $(1 - \delta/2)K < K$, from (32.1.3), there exists $y \in F$ (from which $d_C(y) \geq \eta > 0$) such that

$$\frac{\lambda - g(y)}{d_C(y)} > \left(1 - \frac{\delta}{2}\right) K.$$

Clearing of fractions and using (32.1.5),

$$g(y) + K d_C(y) \leq \lambda + \frac{\delta K d_C(y)}{2} \leq \inf_E (g + K d_C) + \frac{d_C(y)}{2} \delta K.$$

Since d_C is convex and continuous (exercise!), we derive from the Brøndsted–Rockafellar theorem, Corollary 29.2, with $f := g + K d_C$, $\alpha := d_C(y)/2$ and $\beta := \delta K$, that there exists $(x, z^*) \in G(\partial(g + K d_C))$ such that

$$\|x - y\| \leq \frac{d_C(y)}{2} \quad \text{and} \quad \|z^*\| \leq \delta K.$$

Now

$$d_C(y) \le d_C(x) + \|x - y\| \le d_C(x) + \frac{d_C(y)}{2}.$$

Combining this with (32.1.2),

$$d_C(x) \ge \frac{d_C(y)}{2} \ge \frac{\eta}{2} > 0. \tag{32.1.7}$$

From the formula for the subdifferential of a sum, Theorem 28.2, there exist $x^* \in \partial g(x)$ and $y^* \in \partial(Kd_C)(x)$ such that $x^* + y^* = z^*$. Let $w \in C$. Then

$$\langle x - w, x^* \rangle = \langle w - x, y^* \rangle + \langle x - w, z^* \rangle.$$

Since $y^* \in \partial(Kd_C)(x)$ and $\|z^*\| \le \delta K$,

$$\begin{aligned}\langle x - w, x^* \rangle &\le K d_C(w) - K d_C(x) + \delta K \|x - w\| \\ &= K \left(\delta \|x - w\| - d_C(x)\right) \\ &\le K \left(\delta d_C(x) + \delta \operatorname{diam} C - d_C(x)\right).\end{aligned}$$

Thus, from (32.1.6) and (32.1.7),

$$\begin{aligned}\langle x - w, x^* \rangle &< K \left(\frac{1}{2} d_C(x) + \frac{\eta}{4} - d_C(x)\right) \\ &= K \left(\frac{\eta}{4} - \frac{1}{2} d_C(x)\right) \\ &\le K \left(\frac{\eta}{4} - \frac{1}{2} \cdot \frac{\eta}{2}\right) = 0.\end{aligned}$$

This completes the proof of (32.1.1). We now fill in the details of the proof of (32.1.4). We fix $z \in F$. Since g is lower semicontinuous, we can choose $\theta > 0$ such that

$$g(u) \le g(z) - 1 \quad \Longrightarrow \quad \|u - z\| \ge \theta. \tag{32.1.8}$$

Let y be an arbitrary element of F. If $g(y) > g(z) - 1$ then, from (32.1.2),

$$\frac{\lambda - g(y)}{d_C(y)} \le \frac{\lambda - g(z) + 1}{\eta} < \infty.$$

If, on the other hand, $g(y) \le g(z) - 1$ let

$$\gamma := \frac{1}{g(z) - g(y)} \in (0, 1].$$

Then

$$g(\gamma y + (1 - \gamma) z) \le \gamma g(y) + (1 - \gamma) g(z) = g(z) - 1$$

and so, using (32.1.8) with $u := \gamma y + (1 - \gamma) z$,

$$\gamma \|y - z\| = \|\gamma y + (1 - \gamma) z - z\| \ge \theta$$

from which

$$g(z) - g(y) = \frac{1}{\gamma} \le \frac{\|y-z\|}{\theta}.$$

Thus

$$\lambda - g(y) \le \lambda - g(z) + \frac{\|y-z\|}{\theta} \le \lambda - g(z) + \frac{d_C(z) + \operatorname{diam} C + d_C(y)}{\theta}.$$

Consequently,

$$\frac{\lambda - g(y)}{d_C(y)} \le \frac{\lambda - g(z)}{d_C(y)} + \frac{d_C(z) + \operatorname{diam} C}{\theta d_C(y)} + \frac{1}{\theta}$$

and, using (32.1.2) again,

$$\frac{\lambda - g(y)}{d_C(y)} \le \frac{\lambda - g(z)}{\eta} + \frac{d_C(z) + \operatorname{diam} C}{\theta \eta} + \frac{1}{\theta} < \infty.$$

Thus $K < \infty$. This completes the proof of (32.1.4), and hence that of Lemma 32.1. ∎

Remark 32.2. We have given a direct proof of (32.1.4) above. Readers with some knowledge of epigraphical analysis might recognize that it can also be deduced from Beer, [8], Lemma 3.4, p. 277.

Corollary 32.3 can easily be bootstrapped from the contrapositive form of Lemma 32.1 by the substitution $g := f - w^*$.

Corollary 32.3. *Let C be a nonempty $w(E, E^*)$–compact convex subset of E, $w^* \in E^*$, $f \in \mathcal{PCLSC}(E)$ and,*

$$\text{for all } (y, y^*) \in G(\partial f), \ \text{ there exists } w \in C \text{ such that } \langle w - y, w^* - y^* \rangle \ge 0.$$

Then

$$\text{there exists } w \in C \text{ such that } \quad (w, w^*) \in G(\partial f).$$

We now establish the "dual" form of Corollary 32.3, which has a much simpler proof.

Lemma 32.4. *Let C be a nonempty $w(E^*, E)$–compact convex subset of E^*, $w \in E$, $f \in \mathcal{PCLSC}(E)$ and,*

$$\left.\begin{array}{l} \text{for all } (y, y^*) \in G(\partial f), \\ \qquad \text{there exists } w^* \in C \text{ such that } \quad \langle w - y, w^* - y^* \rangle \ge 0. \end{array}\right\} \tag{32.4.1}$$

Then

$$\text{there exists } w^* \in C \text{ such that } \quad (w, w^*) \in G(\partial f). \tag{32.4.2}$$

Proof. We define the sublinear functional S on E by the same formula as in Lemma 30.1, that is to say

$$S(x) := \max\langle x, C\rangle \quad (x \in E).$$

It follows from (32.4.1) that

$$(y, y^*) \in G(\partial f) \quad\Longrightarrow\quad \langle y - w, y^*\rangle + S(w - y) \geq 0. \tag{32.4.3}$$

We also define the continuous convex function $g\colon E \mapsto \mathbb{R}$ by

$$g(y) := S(w - y) \quad (y \in E). \tag{32.4.4}$$

Then (exercise!)

$$y \in E \quad\Longrightarrow\quad \partial g(y) = -\partial S(w - y)$$

hence, from Lemma 29.3,

$$\begin{aligned}(y, z^*) \in G(\partial g) \quad &\Longrightarrow\quad -z^* \in \partial S(w - y)\\ &\Longrightarrow\quad \langle w - y, -z^*\rangle = S(w - y)\\ &\Longrightarrow\quad \langle y - w, z^*\rangle = S(w - y).\end{aligned}$$

Combining this with (32.4.3), we obtain

$$(y, y^*) \in G(\partial f) \text{ and } (y, z^*) \in G(\partial g) \quad\Longrightarrow\quad \langle y - w, y^*\rangle + \langle y - w, z^*\rangle \geq 0,$$

that is to say

$$(y, x^*) \in G(\partial f + \partial g) \quad\Longrightarrow\quad \langle y - w, x^*\rangle \geq 0.$$

Thus, from the formula for the subdifferential of a sum, Theorem 28.2,

$$(y, x^*) \in G(\partial(f + g)) \quad\Longrightarrow\quad \langle y - w, x^* - 0\rangle \geq 0.$$

Since $\partial(f + g)$ is maximal monotone, we derive from this that

$$0 \in \partial(f + g)(w),$$

and so

$$y \in \operatorname{dom} f \quad\Longrightarrow\quad (f + g)(y) - (f + g)(w) \geq 0.$$

Hence, using the definition of g in (32.4.4), we derive that

$$y \in \operatorname{dom} f \quad\Longrightarrow\quad f(y) + S(w - y) - f(w) \geq 0. \tag{32.4.5}$$

Let $A := \operatorname{dom} f$, and define $h\colon A \times C \mapsto \mathbb{R}$ by

$$h(y, w^*) := f(y) + \langle w - y, w^*\rangle - f(w).$$

Then, from (32.4.5) and the definition of S,

$$\inf_A \max_C h \geq 0.$$

The function h is convex on A, and affine and continuous on C. Thus from the minimax theorem, Theorem 3.1,

$$\max_C \inf_A h \geq 0,$$

and so there exists $w^* \in C$ such that

$$y \in \operatorname{dom} f \quad \Longrightarrow \quad f(y) + \langle w - y, w^* \rangle - f(w) \geq 0.$$

This establishes (32.4.2), and completes the proof of Lemma 32.4. ▌

We now come to the main result of this section, which is obtained by combining Corollary 32.3 and Lemma 32.4:

Theorem 32.5. *Let $f \in \mathcal{PCLSC}(E)$. Then $\partial f\colon E \mapsto 2^{E^*}$ is strongly maximal monotone.*

Remark 32.6. The result of Theorem 32.5 lends some credence to the following

Conjecture: If $f \in \mathcal{PCLSC}(E)$, C is a nonempty $w(E, E^*)$–compact convex subset of E, C^* is a nonempty $w(E^*, E)$–compact convex subset of E^* and,

for all $(y, y^*) \in G(\partial f)$,
there exists $(w, w^*) \in C \times C^*$ such that $\langle w - y, w^* - y^* \rangle \geq 0$

then

$$(C \times C^*) \cap G(\partial f) \neq \emptyset.$$

It was proved by Kum in [32], Theorem 2, p. 374–375, Luc in [33], Theorem 2.2, p. 368 and Zagrodny in [61], Theorem 3.1, p. 305 that if $E = \mathbb{R}$ then this conjecture is true.

There are also examples in Luc, [33], p. 368–370 and Zagrodny, [61], Example 3.3, p. 306–307 that if $E = \mathbb{R}^2$ then the conjecture fails (even with f a C^1 function).

Finally, it was proved by Zagrodny in [61], Theorem 4.1, p. 307–308 that if C and C^* satisfy the further condition that there exists $(w_0, w_0^*) \in C \times C^*$ satisfying

$$(w, w^*) \in C \times C^* \Longrightarrow \langle w - w_0, w^* - w_0^* \rangle = 0 \tag{32.6.1}$$

then the conjecture is true and further, by an extremely intricate and ingenious argument, in [61], Theorem 5.2, p. 309–314 that if $E = \mathbb{R}^2$ and the conjecture is true then there exists $(w_0, w_0^*) \in C \times C^*$ satisfying (32.6.1).

33. The biconjugate of a pointwise maximum

We now start working on the results that we will need for our proof in Section 35 that subdifferentials are maximal monotone of type (D). This proof is based ultimately on the formula for the biconjugate of the pointwise maximum of a finite number of functions, which is the main result of this section. (See Theorem 33.3.) What is curious is that we can establish this result without having a simple explicit formula for the *conjugate* of the pointwise maximum. As we shall see in Lemma 33.1(b) and Remark 33.4, we have two such formulae, but they are not simple. The results in this section appear in the paper [25] by Fitzpatrick–Simons.

To simplify the statements of the results that we are about to present, we shall write $\mathcal{CC}(E)$ for the set of all real convex continuous functions on E.

It is easy to see using remark 6.3 that if $f \in \mathcal{PCLSC}(E)$ then $f^* \in \mathcal{PCLSC}(E^*)$. We define the *biconjugate*, f^{**}, of f by

$$f^{**}(x^{**}) := (f^*)^*(x^{**}) \quad (x^{**} \in E^{**}).$$

We will prove in Theorem 33.3 that if $g_0 \in \mathcal{PCLSC}(E)$ and $g_1, \ldots, g_m \in \mathcal{CC}(E)$ then $(g_0 \vee \cdots \vee g_m)^{**} = {g_0}^{**} \vee \cdots \vee {g_m}^{**}$.

We note that Lemma 33.1(a) is essentially the formula for the conjugate of a sum, which was proved by Rockafellar in [45], Theorem 20, p. 56: *if* $f \in \mathcal{PCLSC}(E)$, $g \in \mathcal{CC}(E)$ *and* $w^* \in E^*$ *then*

$$(f+g)^*(w^*) = \min_{u^*,\ v^* \in E^*,\ u^*+v^*=w^*} \big[f^*(u^*) + g^*(v^*)\big].$$

Lemma 33.1(b) might lead one to suspect by analogy that, in this situation, $(f \vee g)^*(w^*)$ is given by the formula

$$\min_{\rho \in [0,1],\ u^*,\ v^* \in E^*,\ \rho u^* + (1-\rho)v^* = w^*} \big[\rho f^*(u^*) + (1-\rho)g^*(v^*)\big], \tag{33.0.1}$$

but this is not necessarily true if $f \notin \mathcal{CC}(E)$. (See Remark 33.4 below for more discussion of this question.)

Lemma 33.1. *Let* $f \in \mathcal{PCLSC}(E)$, $g \in \mathcal{CC}(E)$ *and* $w^* \in E^*$.
(a) Let $\rho,\ \sigma > 0$. *Then there exist* $u^*,\ v^* \in E^*$ *such that*

$$\rho u^* + \sigma v^* = w^* \quad \text{and} \quad \rho f^*(u^*) + \sigma g^*(v^*) \le \sup_{\operatorname{dom} f} \big[w^* - \rho f - \sigma g\big].$$

(b) $(f \vee g)^*(w^*) = \min_{\rho \in [0,1]} \sup_{\operatorname{dom} f} \big[w^* - \rho f - (1-\rho) g\big]$.

Proof. (a) Let $\alpha := \sup_{\operatorname{dom} f} \big[w^* - \rho f - \sigma g\big]$. Since the result is immediate if $\alpha = \infty$, we can and will suppose that $\alpha \in \mathbb{R}$. Then

$$\rho f + \big[\sigma g - w^* + \alpha\big] \geq 0 \quad \text{on} \quad E.$$

and so, from Lemma 28.1, there exists $y^* \in E^*$ such that

$$\begin{aligned}(y, z) \in E \quad &\Longrightarrow \quad \rho f(y) + \sigma g(z) - \langle z, w^*\rangle + \alpha + \langle z - y, y^*\rangle \geq 0 \\ &\Longrightarrow \quad \rho\big[\langle y, y^*/\rho\rangle - f(y)\big] + \sigma\big[\langle z, (w^* - y^*)/\sigma\rangle - g(z)\big] \leq \alpha.\end{aligned}$$

Taking the supremum over y and z,

$$\rho f^*(y^*/\rho) + \sigma g^*((w^* - y^*)/\sigma) \leq \alpha.$$

We now obtain (a) by setting $u^* := y^*/\rho$ and $v^* := (w^* - y^*)/\sigma$.
(b) Since

$$\begin{aligned}(f \vee g)^*(w^*) &= \sup_{x \in \operatorname{dom} f} \big[\langle x, w^*\rangle - (f \vee g)(x)\big] \\ &= \sup_{x \in \operatorname{dom} f} \min_{\rho \in [0,1]} \big[\langle x, w^*\rangle - \rho f(x) - (1 - \rho) g(x)\big],\end{aligned}$$

the result follows from the minimax theorem, Theorem 3.1, with $A := \operatorname{dom} f$ and $B := [0, 1]$. ▌

It is an easy consequence of the definitions that if $f \in \mathcal{PCLSC}(E)$ then $f^{**} \in \mathcal{PCLSC}(E^{**})$ and

$$t^{**} \in E^{**},\ f^{**}(t^{**}) \leq 0 \text{ and } w^* \in E^* \quad \Longrightarrow \quad \langle w^*, t^{**}\rangle \leq f^*(w^*). \quad (33.1.1)$$

Lemma 33.2. *Let $f \in \mathcal{PCLSC}(E)$, $g \in \mathcal{CC}(E)$ and $f^{**}(t^{**}) \vee g^{**}(t^{**}) \leq 0$.*
(a) Let $\rho,\ \sigma > 0$ and $w^* \in E^*$. Then $\langle w^*, t^{**}\rangle \leq \sup_{\operatorname{dom} f} \big[w^* - \rho f - \sigma g\big]$.
(b) Let $w^* \in E^*$. Then $\langle w^*, t^{**}\rangle \leq \sup_{\operatorname{dom} f} \big[w^* - g\big]$.
(c) Let $\rho \in [0, 1]$ and $w^* \in E^*$. Then

$$\langle w^*, t^{**}\rangle \leq \sup_{\operatorname{dom} f} \big[w^* - \rho f - (1 - \rho) g\big] \quad (33.2.1)$$

(d) Let $w^* \in E^*$. Then $\langle w^*, t^{**}\rangle \leq (f \vee g)^*(w^*)$.
(e) $(f \vee g)^{**}(t^{**}) \leq 0$.

Proof. (a) Choose u^* and v^* as in Lemma 33.1(a). Then, from (33.1.1),

$$\langle w^*, t^{**}\rangle = \langle \rho u^* + \sigma v^*, t^{**}\rangle = \rho\langle u^*, t^{**}\rangle + \sigma\langle v^*, t^{**}\rangle \leq \rho f^*(u^*) + \sigma g^*(v^*),$$

and the result follows from Lemma 33.1(a).

(b) Using Remark 6.3, we can fix $x^* \in \operatorname{dom} f^*$. Let $\rho > 0$ be arbitrary. From part (a) with w^* replaced by $\rho x^* + w^*$,

$$\begin{aligned}\rho\langle x^*, t^{**}\rangle + \langle w^*, t^{**}\rangle &= \langle \rho x^* + w^*, t^{**}\rangle \\ &\le \sup_{\operatorname{dom} f} \big[\rho x^* + w^* - \rho f - g\big] \\ &\le \sup_{\operatorname{dom} f} \big[\rho x^* - \rho f\big] + \sup_{\operatorname{dom} f} \big[w^* - g\big] \\ &= \rho f^*(x^*) + \sup_{\operatorname{dom} f} \big[w^* - g\big],\end{aligned}$$

and (b) follows by letting $\rho \to 0$.

(c) If $\rho = 0$ then (33.2.1) follows from (b). If $\rho \in (0, 1)$ then (33.2.1) follows from (a). If, finally, $\rho = 1$ then the right–hand side of (33.2.1) is exactly $f^*(w^*)$, and (33.2.1) follows from (33.1.1).

(d) follows from (c) and Lemma 33.1(b), and (e) is immediate from (d). ∎

Theorem 33.3. *Let $g_0 \in \mathcal{PCLSC}(E)$ and $g_1, \ldots, g_m \in \mathcal{CC}(E)$.*
(a) Let $t^{**} \in E^{**}$. Then $(g_0 \vee g_1)^{**}(t^{**}) \le {g_0}^{**}(t^{**}) \vee {g_1}^{**}(t^{**})$.
(b) $(g_0 \vee g_1)^{**} = {g_0}^{**} \vee {g_1}^{**}$.
(c) $(g_0 \vee \cdots \vee g_m)^{**} = {g_0}^{**} \vee \cdots \vee {g_m}^{**}$.

Proof. (a) Let $\alpha := {g_0}^{**}(t^{**}) \vee {g_1}^{**}(t^{**})$. Since the result is immediate if $\alpha = \infty$, we can and will suppose that $\alpha \in \mathbb{R}$. We now obtain the result by applying Lemma 33.2(e) with $f := g_0 - \alpha$ and $g := g_1 - \alpha$.

(b) Since $g_0 \vee g_1 \ge g_0$ on E, $(g_0 \vee g_1)^{**} \ge {g_0}^{**}$ on E^{**}. Similarly, $(g_0 \vee g_1)^{**} \ge {g_1}^{**}$ on E^{**}, and so $(g_0 \vee g_1)^{**} \ge g_0^{**} \vee {g_1}^{**}$ on E^{**}. The result now follows from (a).

(c) This is immediate from (b) and induction. ∎

Remark 33.4. The following example where $f \in \mathcal{PCLSC}(\mathbb{R}^2)$, $g \in \mathcal{CC}(\mathbb{R}^2)$ but (33.0.1) fails is due to S. Fitzpatrick (personal communication). Define f and g by

$$f(x_1, x_2) := \begin{cases} x_2 & \text{if } x_1 \ge 0; \\ \infty & \text{otherwise;} \end{cases}$$

and

$$g(x_1, x_2) := x_1.$$

Then $(f \vee g)^*(0) = -\inf(f \vee g) = 0$. On the other hand, f^* is the indicator function of $(-\infty, 0] \times \{1\}$ and g^* is the indicator function of $\{(1, 0)\}$. Consequently, if $\rho \in [0, 1]$, $u^* \in \mathbb{R}^2$, $v^* \in \mathbb{R}^2$ and $\rho u^* + (1 - \rho)v^* = 0$ then $\rho f^*(u^*) + (1 - \rho)g^*(v^*) = \infty$, and so (33.0.1) fails.

Here is an explicit formula for $(f \vee g)^*(w^*)$ when $f \in \mathcal{PCLSC}(E)$, $g \in \mathcal{CC}(E)$ and E is a general Banach space. This formula is defined by a two–stage process as follows. If $w^* \in E^*$ and $\delta > 0$, let

$$B(w^*, \delta) := \{x^* \in E^* \colon \|x^* - w^*\| < \delta\}$$

and $L(w^*, \delta)$ be the set

$$\{(\rho, \sigma, u^*, v^*) \colon \rho > 0,\ \sigma > 0,\ u^*,\ v^* \in E^*,\ \rho + \sigma = 1,\ \rho u^* + \sigma v^* \in B(w^*, \delta)\}.$$

If $w^* \in E^*$, write

$$(f^* \underset{\delta}{\wedge} g^*)(w^*) := \inf_{(\rho,\sigma,u^*,v^*) \in L(w^*,\delta)} \left[\rho f^*(u^*) + \sigma g^*(v^*)\right] \quad (\delta > 0)$$

and

$$(f^* \underset{0}{\wedge} g^*)(w^*) := \sup_{\delta > 0} (f^* \underset{\delta}{\wedge} g)^*(w^*) = \lim_{\delta \to 0} (f^* \underset{\delta}{\wedge} g^*)(w^*).$$

Then the formula is that

$$(f \vee g)^*(w^*) = (f^* \underset{0}{\wedge} g^*)(w^*).$$

We refer the reader to [25] for a proof of this fact, in the more general context where $f,\ g \in \mathcal{PCLSC}(E)$, and f and g satisfy the *Attouch–Brézis constraint qualification*

$$\bigcup_{\lambda > 0} \lambda(\operatorname{dom} f - \operatorname{dom} g) \quad \text{is a closed subspace of } E.$$

[25] also contains other results on the binary operation $\underset{0}{\wedge}$.

34. Biconjugate topologies on the bidual

In this section, we continue working towards our proof in Section 35 that subdifferentials are maximal monotone of type (D). We will define a topology $\mathcal{T}_{\mathcal{CLB}}(E^{**})$ on E^{**}. The main result of this section is Theorem 34.7, in which we use Theorem 33.3 to show that if $f \in \mathcal{PCLSC}(E)$ then the canonical image of $G(\partial f)$ is dense in $G(\partial f^*)$ in the topology $\mathcal{T}_{\|\ \|}(E^*) \times \mathcal{T}_{\mathcal{CLB}}(E^{**})$, where "$\mathcal{T}_{\|\ \|}$" stands for "norm topology of".

For comparison purposes, we also introduce a topology $\mathcal{T}_{\mathcal{CC}}(E^{**})$ on E^{**}, which is finer than $\mathcal{T}_{\mathcal{CLB}}(E^{**})$ and has a more "natural" definition. However, we do not know if the approximation property referred to above is true with $\mathcal{T}_{\mathcal{CC}}(E^{**})$ replaced by $\mathcal{T}_{\mathcal{CLB}}(E^{**})$. (See Problem 34.9.)

The portion of this section that will be used to prove that subdifferentials are maximal monotone of type (D) is the part up to and including Theorem 34.7.

Lemma 34.1. *Let $f_0 \in \mathcal{PCLSC}(E)$ and $f_1, \ldots, f_m \in \mathcal{CC}(E)$. Let $t^{**} \in E^{**}$ and $\delta > 0$. Then there exists $t \in E$ such that*

$$\text{for all } i = 0, \ldots, m, \quad f_i(t) \leq f_i^{**}(t^{**}) + \delta.$$

Proof. Since we can remove those values of i for which $f_i^{**}(t^{**}) = \infty$, we can and will suppose that $f_0^{**}(t^{**}), \ldots, f_m^{**}(t^{**}) \in \mathbb{R}$. For all $i = 0, \ldots, m$, let $g_i := f_i - f_i^{**}(t^{**})$. Then $g_i^{**}(t^{**}) = 0$, hence $g_0^{**}(t^{**}) \vee \cdots \vee g_m^{**}(t^{**}) = 0$. From Theorem 33.3(c), $(g_0 \vee \cdots \vee g_m)^{**}(t^{**}) = 0$ and so, from (33.1.1) with $f := g_0 \vee \cdots \vee g_m$ and $w^* := 0$, $(g_0 \vee \cdots \vee g_m)^*(0) \geq 0$, that is to say $\inf_E(g_0 \vee \cdots \vee g_m) \leq 0$ The result follows by rewriting this inequality in terms of the functions f_i. ∎

Definition 34.2. We write $\mathcal{CLB}(E)$ for the set of all convex functions f: $E \mapsto \mathbb{R}$ that are Lipschitz on the bounded subsets of E, or equivalently (exercise!) bounded on the bounded subsets of E. The standard example of a function $f \in \mathcal{CC}(\ell^2) \setminus \mathcal{CLB}(\ell^2)$ is defined by $f(x) := \sum_{n=1}^{\infty} n x_n{}^{2n}$ ($x = \{x_n\}_{n \geq 1} \in \ell^2$). It was proved by Borwein–Fitzpatrick–Vanderwerff in [11], Theorem 2.2, p. 64 using the deep Josefson-Nissenzweig theorem that if E is infinite dimensional then $\mathcal{CC}(E) \setminus \mathcal{CLB}(E) \neq \emptyset$.

Theorem 34.3. *Let $f \in \mathcal{PCLSC}(E)$, and $(t^*, t^{**}) \in G(\partial f^*)$. Let $m \geq 1$, $f_1, \ldots, f_m \in \mathcal{CLB}(E)$ and $\varepsilon > 0$. Then there exists $(s, s^*) \in G(\partial f)$ such that*

$$\|s^* - t^*\| \leq \varepsilon \tag{34.3.1}$$

$$f(s) \leq f^{**}(t^{**}) + \varepsilon \tag{34.3.2}$$

and

$$\text{for all } i = 1, \ldots, m, \quad f_i(s) \leq f_i^{**}(t^{**}) + \varepsilon. \tag{34.3.3}$$

Proof. Define f_{m+1}, $f_{m+2} \in \mathcal{CLB}(E)$ by $f_{m+1} := \|\ \|$ and $f_{m+2} := t^*$, and let L be the maximum of the Lipschitz constants of $f_1, \ldots f_{m+2}$ on the set $\{x \in E\colon \|x\| \leq \|t^{**}\| + 2\}$. Choose $\delta \in (0, 1]$ such that

$$2\delta + \sqrt{\delta} L \leq \varepsilon \tag{34.3.4}$$

Finally, define f_0: $E \mapsto \mathbb{R} \cup \{\infty\}$ by $f_0 := f - t^*$. From Lemma 34.1 with m replaced by $m + 2$, there exists $t \in E$ such that,

$$\text{for all } i = 0, \ldots, m + 2, \quad f_i(t) \leq f_i^{**}(t^{**}) + \delta. \tag{34.3.5}$$

When $i = 0$ and $i = m + 2$, (34.3.5) yields

$$f(t) - \langle t, t^* \rangle \leq f^{**}(t^{**}) - \langle t^*, t^{**} \rangle + \delta \quad \text{and} \quad \langle t, t^* \rangle \leq \langle t^*, t^{**} \rangle + \delta$$

thus, adding these two inequalities,

$$f(t) \leq f^{**}(t^{**}) + 2\delta. \tag{34.3.6}$$

On the other hand, since $t^{**} \in \partial f^*(t^*)$, $f_0^{**}(t^{**}) = -f_0^*(0) = \inf_E f_0$ and so, when $i = 0$, (34.3.5) also yields $f_0(t) \le \inf_E f_0 + \delta$. It follows from the Brøndsted-Rockafellar theorem, Corollary 29.2, that there exists $(s, z^*) \in G(\partial f_0)$ such that

$$\|z^*\| \le \sqrt{\delta}, \tag{34.3.7}$$

$$f_0(s) \le f_0(t) \tag{34.3.8}$$

and

$$\|s - t\| \le \sqrt{\delta} \le 1. \tag{34.3.9}$$

Let $s^* := t^* + z^*$. Since $(s, z^*) \in G(\partial f_0)$, $(s, s^*) \in G(\partial f)$, as required. It is clear by considering f_{m+1} that $L \ge 1$ thus, from (34.3.4), $\sqrt{\delta} \le \varepsilon$. Since $s^* - t^* = z^*$, (34.3.1) now follows from (34.3.7). Further, from (34.3.8), (34.3.6) and (34.3.4),

$$f(s) \le f(t) + \langle s - t, t^* \rangle \le f^{**}(t^{**}) + 2\delta + \sqrt{\delta}L \le f^{**}(t^{**}) + \varepsilon,$$

which gives (34.3.2). Finally, we establish (34.3.3). When $i = m + 1$, (34.3.5) yields

$$\|t\| \le \|t^{**}\| + \delta \le \|t^{**}\| + 1,$$

thus, from (34.3.9),

$$\|s\| \le \|s - t\| + \|t\| \le \|t^{**}\| + 2.$$

It now follows from (34.3.9) again and the choice of L that

$$\text{for all } i = 1, \ldots, m, \quad |f_i(s) - f_i(t)| \le \sqrt{\delta}L$$

and we obtain (34.3.3) by combining this with (34.3.5) and (34.3.4). This completes the proof of Theorem 34.3. ▌

Remark 34.4. Let $f \in \mathcal{PCLSC}(E)$. It is immediate from the definitions and Remark 6.3 that

$$f^{**} \text{ is } w(E^{**}, E^*)\text{–lower semicontinuous} \tag{34.4.1}$$

and

$$x \in E \implies f^{**}(\widehat{x}) = f(x). \tag{34.4.2}$$

It also follows (exercise!) that if $x^{**} \in E^{**}$ then

$$f^{**}(x^{**}) \le \sup\{f(x)\colon x \in E,\ \|x\| \le \|x^{**}\|\}.$$

Consequently, if $f \in \mathcal{CLB}(E)$ then $f^{**} \in \mathcal{CLB}(E^{**})$. (See, for instance, Phelps, [34], Proposition 3.3, p. 39 for hints on how to prove that f^{**} is continuous.) In general, $\mathcal{CC}(E)$ behaves in a much more pathological fashion. To illustrate this, we give an example of a function $f \in \mathcal{CC}(c_0)$ such that f^{**} is not continuous on $c_0{}^{**} = \ell^\infty$. Define $f \in \mathcal{CC}(c_0)$ by

$$f(x) := \sum_{n\geq 1} \frac{1}{n^2} x_n{}^{2n} \quad (x = \{x_n\}_{n\geq 1} \in c_0).$$

Then, by direct computation,

$$f^{**}(x^{**}) = \sum_{n\geq 1} \frac{1}{n^2} \xi_n{}^{2n} \quad (x^{**} = \{\xi_n\}_{n\geq 1} \in \ell^\infty).$$

Now, for $N \geq 1$, define $\xi^N \in \ell^\infty$ by

$$\xi^N{}_n := \begin{cases} 1, & \text{if } n \leq N; \\ \sqrt[n]{n}, & \text{otherwise.} \end{cases}$$

Since $\lim_{n\to\infty} \sqrt[n]{n} = 1$, $\xi^N \to (1, 1, \ldots)$ in ℓ^∞ as $N \to \infty$. On the other hand

$$f^{**}(\xi^N) = \sum_{n\leq N} \frac{1}{n^2} + \sum_{n>N} n^2 \frac{1}{n^2} = \infty$$

and

$$f^{**}((1, 1, \ldots)) = \sum_{n\geq 1} \frac{1}{n^2} < \infty.$$

So $f^{**}(\xi^N) \not\to f^{**}((1, 1, \ldots))$ as $N \to \infty$. Consequently, f^{**} is not continuous on ℓ^∞.

Definition 34.5. We define the topology $\mathcal{T}_{\mathcal{CC}}(E^{**})$ on E^{**} to be the coarsest topology on E^{**} making all the functions $f^{**} : E^{**} \mapsto \mathbb{R}\cup\{\infty\} \quad (f \in \mathcal{CC}(E))$ continuous. We define the topology $\mathcal{T}_{\mathcal{CLB}}(E^{**})$ on E^{**} to be the coarsest topology on E^{**} making all the functions $f^{**} : E^{**} \mapsto \mathbb{R} \quad (f \in \mathcal{CLB}(E))$ continuous.

It is clear from Remark 34.4 that $\mathcal{T}_{\mathcal{CLB}}(E^{**}) \subset \mathcal{T}_{\|\ \|}(E^{**})$, and that it is not true in general that $\mathcal{T}_{\mathcal{CC}}(E^{**}) \subset \mathcal{T}_{\|\ \|}(E^{**})$. It also follows from Lemma 34.6(a) below that $w(E^{**}, E^*) \subset \mathcal{T}_{\mathcal{CLB}}(E^{**}) \subset \mathcal{T}_{\mathcal{CC}}(E^{**})$.

Lemma 34.6. *Let $\{x_\alpha^{**}\}$ be a net of elements of E^{**} and $x^{**} \in E^{**}$.*
(a) If $x_\alpha^{**} \to x^{**}$ in $\mathcal{T}_{\mathcal{CLB}}(E^{**})$ then $x_\alpha^{**} \to x^{**}$ in $w(E^{**}, E^*)$.
(b) If $x_\alpha^{**} \to x^{**}$ in $\mathcal{T}_{\mathcal{CLB}}(E^{**})$ then $\{x_\alpha^{**}\}$ is eventually bounded in the norm of E^{**}.
(c) $x_\alpha^{**} \to x^{**}$ in $\mathcal{T}_{\mathcal{CLB}}(E^{**})$ if, and only if,

$$\text{for all } f \in \mathcal{CLB}(E), \quad \limsup_\alpha f^{**}(x_\alpha^{**}) \leq f^{**}(x^{**}). \tag{34.6.1}$$

(d) $x_\alpha^{**} \to x^{**}$ in $\mathcal{T}_{\mathcal{CC}}(E^{**})$ if, and only if,

$$\text{for all } f \in \mathcal{CC}(E), \quad \limsup_\alpha f^{**}(x_\alpha^{**}) \leq f^{**}(x^{**}).$$

Proof. (a) is immediate since $x^* \in E^*$ implies that $x^* \in \mathcal{CLB}(E)$ and $-x^* \in \mathcal{CLB}(E)$, and (b) is immediate since $\| \; \| \in \mathcal{CLB}(E)$. "Only if" in (c) is also immediate. Suppose, conversely, that (34.6.1) is true. Let $x^* \in E^*$. Then, since $x^* \in \mathcal{CLB}(E)$ and $-x^* \in \mathcal{CLB}(E)$, we have from (34.6.1) that

$$\limsup_\alpha \langle x^*, x_\alpha^{**}\rangle \le \langle x^*, x^{**}\rangle \quad \text{and} \quad \liminf_\alpha \langle x^*, x_\alpha^{**}\rangle \ge \langle x^*, x^{**}\rangle,$$

thus $x_\alpha^{**} \to x^{**}$ in $w(E^{**}, E^*)$. From (34.4.1),

$$\text{for all } f \in \mathcal{CLB}(E), \quad \liminf_\alpha f^{**}(x_\alpha^{**}) \ge f^{**}(x^{**}).$$

Combining this with (34.6.1),

$$\text{for all } f \in \mathcal{CLB}(E), \quad f^{**}(x_\alpha^{**}) \to f^{**}(x^{**}),$$

that is to say, $x_\alpha^{**} \to x^{**}$ in $\mathcal{T}_{\mathcal{CLB}}(E^{**})$. This completes the proof of "if" of (c). The proof of (d) is similar to that of (c). ∎

We now come to the main result of this section, Theorem 34.7, which we will use in the next section to prove that subdifferentials are maximal monotone of type (D). Theorems 34.7 and 34.8 represent further sharpenings of results proved by Rockafellar in [44], Proposition 1, p. 211 and subsequently sharpened by Gossez in [26], Théorème 3.1 and Lemme 3.1, p. 376–378.

Theorem 34.7. *Let $f \in \mathcal{PCLSC}(E)$, and $(t^*, t^{**}) \in G(\partial f^*)$. Then there exists a net $\{(s_\alpha, s_\alpha^*)\}$ of elements of $G(\partial f)$ such that $\widehat{s_\alpha} \to t^{**}$ in $\mathcal{T}_{\mathcal{CLB}}(E^{**})$, $s_\alpha^* \to t^*$ in norm and $f(s_\alpha) \to f^{**}(t^{**})$.*

Proof. It is clear from Theorem 34.3 that there exists a net $\{(s_\alpha, s_\alpha^*)\}$ of elements of $G(\partial f)$ such that $s_\alpha^* \to t^*$ in norm,

$$\limsup_\alpha f(s_\alpha) \le f^{**}(t^{**}) \tag{34.7.1}$$

and

$$\text{for all } g \in \mathcal{CLB}(E), \quad \limsup_\alpha g(s_\alpha) \le g^{**}(t^{**}).$$

(We can take as the indexing set of the net the cartesian product of $(0, \infty)$ and the set of nonempty finite subsets of $\mathcal{CLB}(E)$.) It follows from (34.4.2) that

$$\text{for all } g \in \mathcal{CLB}(E), \quad \limsup_\alpha g^{**}(\widehat{s_\alpha}) \le g^{**}(t^{**})$$

and so, using Lemma 34.6(c), $\widehat{s_\alpha} \to t^{**}$ in $\mathcal{T}_{\mathcal{CLB}}(E^{**})$, as required. From Lemma 34.6(a), $\widehat{s_\alpha} \to t^{**}$ in $w(E^{**}, E^*)$ and so, from (34.4.1) and (34.4.2),

$$f^{**}(t^{**}) \le \liminf_\alpha f^{**}(\widehat{s_\alpha}) = \liminf_\alpha f(s_\alpha).$$

It follows by combining this with (34.7.1) that $f(s_\alpha) \to f^{**}(t^{**})$. ∎

Theorem 34.8. *Let $f \in \mathcal{PCLSC}(E)$, and $t^{**} \in E^{**}$. Then there exists a net $\{t_\alpha\}$ of elements of E such that $\widehat{t_\alpha} \to t^{**}$ in $\mathcal{T}_{\mathcal{CC}}(E^{**})$ and $f(t_\alpha) \to f^{**}(t^{**})$.*

Proof. This follows exactly the same lines as the proof of Theorem 34.7, except that it uses Lemma 34.1 instead of Theorem 34.3. ▌

A comparison of Theorems 34.7 and 34.8 leads naturally to the following question:

Problem 34.9. Let $f \in \mathcal{PCLSC}(E)$, and $(t^*, t^{**}) \in G(\partial f^*)$. Does there necessarily exist a net $\{(s_\alpha, s_\alpha^*)\}$ of elements of $G(\partial f)$ such that $\widehat{s_\alpha} \to t^{**}$ in $\mathcal{T}_{\mathcal{CC}}(E^{**})$, $s_\alpha^* \to t^*$ in norm and $f(s_\alpha) \to f^{**}(t^{**})$? In other words, can we replace "$\mathcal{T}_{\mathcal{CLB}}(E^{**})$" by "$\mathcal{T}_{\mathcal{CC}}(E^{**})$" in the statement of Theorem 34.7? (If we could, then it would be unnecessary to introduce the topology $\mathcal{T}_{\mathcal{CLB}}(E^{**})$ or even the set of functions $\mathcal{CLB}(E)$.)

We now investigate the topology $\mathcal{T}_{\mathcal{CLB}}(E^{**})$ more closely.

Theorem 34.10. (a) *Let $\{x_\alpha\}$ be a net of elements of E and $x \in E$. Then*

$$\widehat{x_\alpha} \to \widehat{x} \text{ in } \mathcal{T}_{\mathcal{CLB}}(E^{**}) \iff x_\alpha \to x \text{ in } \mathcal{T}_{\|\ \|}(E).$$

(b) $\mathcal{T}_{\mathcal{CLB}}(E^{**}) = w(E^{**}, E^*) \iff E$ is finite dimensional.
(c) $\mathcal{T}_{\mathcal{CLB}}(E^{**}) = \mathcal{T}_{\|\ \|}(E^{**}) \iff E$ is reflexive.

Proof. (a) We leave this as an exercise. (For ($\Longrightarrow$) consider the element f of $\mathcal{CLB}(E)$ defined by $f(y) := \|y - x\| \quad (y \in E)$.)

(b)($\Longrightarrow$) If $\mathcal{T}_{\mathcal{CLB}}(E^{**}) = w(E^{**}, E^*)$ then, from (a), $\mathcal{T}_{\|\ \|}(E) = w(E, E^*)$. It is well known that this implies that E is finite dimensional. ($\Longleftarrow$) follows since, as we have already observed, $w(E^{**}, E^*) \subset \mathcal{T}_{\mathcal{CLB}}(E^{**}) \subset \mathcal{T}_{\|\ \|}(E^{**})$.

(c)($\Longrightarrow$) Let t^{**} be an arbitrary element of E^{**}. From Theorem 34.8 with any $f \in \mathcal{PCLSC}(E)$, there exists a net $\{t_\alpha\}$ of elements of E such that $\widehat{t_\alpha} \to t^{**}$ in $\mathcal{T}_{\mathcal{CLB}}(E^{**})$. So if $\mathcal{T}_{\mathcal{CLB}}(E^{**}) = \mathcal{T}_{\|\ \|}(E^{**})$, $\widehat{t_\alpha} \to t^{**}$ in $\mathcal{T}_{\|\ \|}(E^{**})$. Since $\widehat{E}$ is closed in $(E^{**}, \mathcal{T}_{\|\ \|}(E^{**}))$, $t^{**} \in \widehat{E}$. Thus E is reflexive. ($\Longleftarrow$) follows from (a). ▌

Remark 34.11. Our knowledge of $\mathcal{T}_{\mathcal{CLB}}(E^{**})$ is very sketchy. For instance, we do not know whether, if $y^{**} \in E^{**}$,

$$x_\alpha^{**} \to x^{**} \text{ in } (E^{**}, \mathcal{T}_{\mathcal{CLB}}(E^{**})) \implies x_\alpha^{**} + y^{**} \to x^{**} + y^{**} \text{ in } (E^{**}, \mathcal{T}_{\mathcal{CLB}}(E^{**})).$$

On the other hand, we can prove that if $y \in E$ then

$$x_\alpha^{**} \to x^{**} \text{ in } (E^{**}, \mathcal{T}_{\mathcal{CLB}}(E^{**})) \implies \|x_\alpha^{**} - \widehat{y}\| \to \|x^{**} - \widehat{y}\|.$$

In fact, we can prove even more generally that if Y is a nonempty $w(E, E^*)$–compact convex subset of E then

$$x_\alpha^{**} \to x^{**} \text{ in } (E^{**}, \mathcal{T}_{\mathcal{CLB}}(E^{**})) \implies \inf_{y \in Y} \|x_\alpha^{**} - \widehat{y}\| \to \inf_{y \in Y} \|x^{**} - \widehat{y}\|.$$

35. Subdifferentials are maximal monotone of type (D), and more

We shall prove in Theorem 35.3 below the result already advertised in Section 25 that *subdifferentials are maximal monotone of type (D).*

In fact, we will do somewhat more than that. We will define a new class of multifunctions, those that are maximal monotone of type (DS), and what we will actually prove is that *subdifferentials are maximal monotone of type (DS).*

Now it is clear from Lemma 34.6 that every maximal monotone multifunction of type (DS) is automatically of type (D). (The additional "S" stands for "strongly"). In fact, every maximal monotone multifunction of type (DS) is automatically of "dense type" in the sense introduced by Gossez in [26], p. 375. Thus Theorem 35.3 extends the result proved in [26], Théorème 3.1, p. 376–378 that subdifferentials are maximal monotone of dense type.

Definition 35.1. S is *maximal monotone of type (DS)* if S is maximal monotone and, for all $(x^{**}, x^*) \in G(\overline{S})$, there exists a net (w_α, w_α^*) of elements of $G(S)$ such that $\widehat{w_\alpha} \to x^{**}$ in $\mathcal{T}_{\mathcal{CLB}}(E^{**})$ and $\|w_\alpha^* - x^*\| \to 0$.

Lemma 35.2. *Let $f \in \mathcal{PCLSC}(E)$ and $(x^*, x^{**}) \in E^* \times E^{**}$. Then the conditions* (35.2.1)–(35.2.3) *are equivalent:*

$$(x^{**}, x^*) \in G(\overline{\partial f}) \tag{35.2.1}$$

$$\inf_{(t^*, t^{**}) \in G(\partial f^*)} \langle t^* - x^*, t^{**} - x^{**} \rangle \geq 0 \tag{35.2.2}$$

$$(x^*, x^{**}) \in G(\partial f^*). \tag{35.2.3}$$

Proof. ((35.2.1)$\Longrightarrow$(35.2.2)) Let (t^*, t^{**}) be an arbitrary element of $G(\partial f^*)$. From Theorem 34.7 and Lemma 34.6, there exists a net $\{(s_\alpha, s_\alpha^*)\}$ of elements of $G(\partial f)$ such that $\widehat{s_\alpha}$ is eventually bounded, $\widehat{s_\alpha} \to t^{**}$ in $w(E^{**}, E^*)$ and $s_\alpha^* \to t^*$ in norm. From (35.2.1),

$$\text{for all } \alpha, \quad \langle s_\alpha^* - x^*, \widehat{s_\alpha} - x^{**} \rangle \geq 0$$

hence, by passing to the limit,

$$\langle t^* - x^*, t^{**} - x^{**} \rangle \geq 0.$$

Thus we have established (35.2.2).
((35.2.2)$\Longrightarrow$(35.2.1)) This is immediate since, from (34.4.2),

$$(s, s^*) \in G(\partial f) \iff (s^*, \widehat{s}) \in G(\partial f^*).$$

((35.2.2)$\Longleftrightarrow$(35.2.3)) This equivalence follows since, from Rockafellar's maximal monotonicity theorem, $\partial f^*\colon E^* \mapsto 2^{E^{**}}$ is maximal monotone. ▌

Theorem 35.3. *Let $f \in \mathcal{PCLSC}(E)$. Then $\partial f\colon E \mapsto 2^{E^*}$ is maximal monotone of type (DS).*

Proof. This follows from Lemma 35.2 and Theorem 34.7. ▌

VIII. Discontinuous positive linear operators

36. A criterion for maximality

Let $D(T)$ be a linear subspace of E and $T\colon D(T) \mapsto E^*$ a (possibly unbounded) linear operator such that

$$s \in D(T) \quad \Longrightarrow \quad \langle s, Ts \rangle \geq 0.$$

In Section 8, we observed that if $D(T) = E$ then T is maximal monotone. Furthermore, T is automatically continuous (exercise!). In this section, we consider the situation when $D(T)$ is a *proper* subspace of E. It is still true and easy to see that T is monotone. In Theorem 36.2, we shall use the minimax theorem to find a necessary and sufficient condition for T to be maximal monotone. This criterion is in terms of the set $H(T)$ defined below. ($H(T)$ stands for the *halo* of T.) We note that $H(T)$ is closed under multiplication by scalars and also that $H(T)$ is an F_σ. The results in this section appear in the paper [36] by Phelps–Simons, with proofs based on the Eidelheit separation theorem rather than a minimax theorem.

Lemma 36.1. *Let $T\colon D(T) \mapsto E^*$ be monotone and linear. Write*

$$\left. \begin{aligned} H(T) := \{z \in E\colon & \text{ there exists } M \geq 0 \text{ such that} \\ & y \in D(T) \Longrightarrow \langle z - y, Ty \rangle \leq M\|z - y\|\}. \end{aligned} \right\} \qquad (36.1.1)$$

Let $z \in E$. Then:

$$z \in H(T) \iff \text{there exists } z^* \in E^* \text{ such that } (z, z^*) \text{ is monotonically related to } G(T).$$

Proof. We have $z \in H(T) \iff$

$$\text{there exists } M \geq 0 \text{ such that} \quad y \in D(T) \Longrightarrow M\|z - y\| - \langle z - y, Ty \rangle \geq 0.$$

From the one–dimensional Hahn–Banach theorem, Corollary 1.2, this can be written

$$\left. \begin{aligned} & \text{there exists } M \geq 0 \text{ such that} \\ & \quad y \in D(T) \Longrightarrow \max_{\|z^*\| \leq M} \langle z - y, z^* \rangle - \langle z - y, Ty \rangle \geq 0, \end{aligned} \right\}$$

that is to say,

$$\text{there exists } M \geq 0 \text{ such that } \quad y \in D(T) \Longrightarrow \max_{\|z^*\| \leq M} \langle y - z, Ty - z^* \rangle \geq 0,$$

which is equivalent to

$$\text{there exists } M \geq 0 \text{ such that } \quad \inf_{y \in D(T)} \max_{\|z^*\| \leq M} \langle y - z, Ty - z^* \rangle \geq 0. \quad (36.1.2)$$

Let $A := D(T)$ and $B := \{z^* \in E^*\colon \|z^*\| \leq M\}$ with the topology $w(E^*, E)$. From the Banach–Alaoglu theorem, Theorem 4.1, B is compact. Define $h\colon A \times B \mapsto \mathbb{R}$ by

$$h(y, z^*) := \langle y - z, Ty - z^* \rangle.$$

The function h is convex on A (exercise!), and affine and continuous on B. Thus from the minimax theorem, Theorem 3.1, (36.1.2) is equivalent to

$$\text{there exists } M \geq 0 \text{ such that } \quad \max_{\|z^*\| \leq M} \inf_{y \in D(T)} \langle y - z, Ty - z^* \rangle \geq 0,$$

that is to say,

$$\text{there exist } M \geq 0 \text{ and } \|z^*\| \leq M \text{ such that } \quad \inf_{y \in D(T)} \langle y - z, Ty - z^* \rangle \geq 0,$$

which is equivalent to

$$\text{there exists } z^* \in E^* \text{ such that } \quad y \in D(T) \Longrightarrow \langle y - z, Ty - z^* \rangle \geq 0.$$

This gives the required result. ∎

Theorem 36.2. *Let $T\colon D(T) \mapsto E^*$ be monotone and linear. Then T is maximal monotone if, and only if, $D(T)$ is dense and $H(T) = D(T)$.*

Proof. ($\Longrightarrow$) Suppose that T is maximal monotone. We shall prove first that

$$z^* \in E^* \text{ and } \langle y, z^* \rangle = 0 \text{ for all } y \in D(T) \quad \Longrightarrow \quad z^* = 0,$$

this will establish that $D(T)$ is dense (see Corollary 4.7). Let $z^* \in E^*$ and $\langle y, z^* \rangle = 0$ for all $y \in D(T)$. Then

$$y \in D(T) \quad \Longrightarrow \quad \langle y - 0, Ty - z^* \rangle = \langle y, Ty - z^* \rangle = \langle y, Ty \rangle \geq 0,$$

thus $(0, z^*)$ is monotonically related to $G(T)$. Since T is maximal, $(0, z^*) \in G(T)$ and so $z^* = T0 = 0$. This completes the proof that $D(T)$ is dense. If $z \in D(T)$ then (z, Tz) is monotonically related to $G(T)$ hence, from Lemma 36.1($\Longleftarrow$), $z \in H(T)$. If, on the other hand, $z \in H(T)$ then, from Lemma 36.1($\Longrightarrow$), there exists $z^* \in E^*$ such that (z, z^*) is monotonically related to $G(T)$. Since T is maximal monotone, $(z, z^*) \in G(T)$, hence $z \in D(T)$. This completes the proof that $H(T) = D(T)$.

($\Longleftarrow$) Let $D(T)$ be dense and $H(T) = D(T)$. Suppose that (z, z^*) is monotonically related to $G(T)$. From Lemma 36.1($\Longleftarrow$), $z \in H(T)$, hence $z \in D(T)$. Now let $x \in D(T)$, and λ be an arbitrary real number. Then $z + \lambda x \in D(T)$, from which (exercise!) $\langle x, Tz - z^* \rangle = 0$. Since this holds for all $x \in D(T)$ and $D(T)$ is dense, $Tz - z^* = 0 \in E^*$, hence $z^* = Tz$. Thus $(z, z^*) = (z, Tz) \in G(T)$. This completes the proof that T is maximal monotone. ▌

37. A sum theorem

If S and T are maximal monotone and linear and Rockafellar's constraint qualification (20.0.1) is satisfied then $\operatorname{int} D(S) \neq \emptyset$ hence, since $D(S)$ is a subspace of E, $D(S) = E$. Thus it follows from Theorem 37.1 that, even *without the assumption of reflexivity*, $S + T$ is maximal monotone in the linear case under condition (20.0.1). Theorem 37.1 appears in the paper [36] by Phelps–Simons, with a proof based on the Eidelheit separation theorem.

Theorem 37.1. *Let S: $E \mapsto E^*$ be monotone and linear and T: $D(T) \mapsto E^*$ be maximal monotone and linear. Then $S + T$: $D(T) \mapsto E^*$ is maximal monotone.*

Proof. Let (x, x^*) be monotonically related to $G(S + T)$. Define convex functions f: $E \mapsto \mathbb{R} \cup \{\infty\}$ and g: $E \mapsto \mathbb{R}$ by

$$f(y) := \begin{cases} \langle y - x, Ty \rangle, & \text{if } y \in D(T); \\ \infty, & \text{otherwise}; \end{cases}$$

and

$$g(z) := \langle z - x, Sz - x^* \rangle \quad (z \in E).$$

Then

$$u \in E \quad \Longrightarrow \quad f(u) + g(u) \geq \langle u - x, (S + T)u - x^* \rangle \geq 0. \tag{37.1.1}$$

Since g is continuous (exercise!), it follows from Lemma 28.1 that there exists $y^* \in E^*$ such that

$$z \in \operatorname{dom} g \text{ and } y \in \operatorname{dom} f \quad \Longrightarrow \quad g(z) + f(y) + \langle z - y, y^* \rangle \geq 0.$$

Substituting in the formulae for f and g, we derive from this that

$$z \in E \text{ and } y \in D(T) \quad \Longrightarrow \quad \langle z - x, Sz - (x^* - y^*) \rangle + \langle y - x, Ty - y^* \rangle \geq 0$$

hence

$$\inf_{z \in E} \langle z - x, Sz - (x^* - y^*) \rangle + \inf_{y \in D(T)} \langle y - x, Ty - y^* \rangle \geq 0.$$

Since S and T are maximal monotone, from Lemma 8.1(c), both the infima above are ≤ 0, thus

$$\inf_{z\in E}\langle z-x, Sz-(x^*-y^*)\rangle \geq 0 \quad \text{and} \quad \inf_{y\in D(T)}\langle y-x, Ty-y^*\rangle \geq 0.$$

Since S and T are maximal monotone, $(x, x^*-y^*) \in G(S)$ and $(x, y^*) \in G(T)$. Consequently, $(x, x^*) = (x, (x^*-y^*)+y^*) \in G(S+T)$. This completes the proof that $S+T$ is maximal monotone. ▌

Problem 37.2. If S and T are maximal monotone and linear and

$$D(S) - D(T) \quad \text{is absorbing} \tag{23.1.2}$$

(or, equivalently, $D(S) - D(T) = E$) then is $S+T$ necessarily maximal monotone? We shall give an example below showing that, even for linear operators in the space ℓ^2, we cannot weaken (23.1.2) further to the condition

$$D(S) - D(T) \quad \text{is dense.}$$

The following examples of discontinuous maximal monotone linear operators are taken from the paper [36] by Phelps–Simons, to which we refer the reader for a more comprehensive analysis.

- Let $E := L^1[0,1]$,

$$D(T) = \{x \in L^1\colon\ x \text{ is Lipschitz and } x(0) = 0\}.$$

Define $T\colon D(T) \mapsto L^\infty$ by $Tx := x'$. Then T is maximal monotone.

- Let $E := L^1[0,1]$,

$$D(T) = \{x \in L^1\colon\ x \text{ is Lipschitz and } x(0) = x(1)\}.$$

Define $T\colon D(T) \mapsto L^\infty$ by $Tx := x'$. Then T is maximal monotone.

- Let $E := \ell^2$, and define V, W: $\ell^2 \mapsto \ell^2$ by

$$Vx := (x_1, x_2 - x_1, x_3 - x_2, \ldots) \quad \text{and} \quad Wx := (x_1 - x_2, x_2 - x_3, x_3 - x_4, \ldots)$$

for $x = \{x_n\}_{n\geq 1} \in \ell^2$. Both V and W are injective, so we can define S and T by $S := V^{-1}$ (with $D(S) = R(V)$) and $T := W^{-1}$ (with $D(T) = R(W)$). S and T are maximal monotone. Even though $D(S) - D(T)$ is dense in ℓ^2, $S+T$ is not maximal monotone (exercise!). Compare this example with Theorem 21.3. (In fact, it is even true that $D(S) \cap D(T)$ is dense in ℓ^2 — see [36].)

38. Discontinuous positive linear operators and the "six subclasses"

In this section, we consider the six subclasses of the maximal monotone multifunctions introduced in Section 25, with reference to the *positive linear* operators.

We first show in Theorem 38.2 that there is no point in looking among the discontinuous positive linear operators for a maximal monotone multifunction that is not of type (FPV), that is to say *every linear maximal monotone operator is of type (FPV).*

It was proved in Bauschke–Borwein, [6], Theorem 4.1, and reproved in a simpler fashion in Phelps–Simons, [36], that *if* $T\colon E \mapsto E^*$ *is continuous, linear and positive then*

$$T \text{ is of type (NI)} \iff T \text{ is of type (FP)}. \tag{38.0.1}$$

In fact, there are many other equivalent conditions in [6], Theorem 4.1, for instance that T^* be positive. However, in this section we are primarily interested in the *discontinuous* linear case. We shall show in Theorem 38.3 below that the implication ($\Longrightarrow$) in (38.0.1) remains true in the discontinuous case. The proof of Theorem 38.3 is similar to that of Theorem 38.2, but is somewhat more technical since it has the added complication of "going into the bidual". Theorems 38.2 and 38.3 appear in [36], with proofs based on the Eidelheit separation theorem rather than a minimax theorem. (We point out that the "decomposition technique" used in [6] does not seem to be applicable to the discontinuous case.)

What *is* true in the discontinuous linear case (and proved in Phelps–Simons, [36]) is that that there is no point in looking among the discontinuous positive linear operators for a solution to Problem 25.7, that is to say, *if* $T\colon D(T) \mapsto E^*$ *is linear and positive then*

$$T \text{ is of type (NI)} \iff T \text{ is of type (D)}.$$

This suggests the following problem:

Problem 38.1. If $T\colon D(T) \mapsto E^*$ is linear and maximal monotone of type (NI), does it automatically follow that T is of type (DS)?

We prove in Theorem 38.5 that there is also no point in looking among the discontinuous positive linear operators for an example of a maximal monotone multifunction that is not strongly maximal monotone. Finally, we prove in Theorem 38.6 that there is no point in looking among the continuous linear operators for an example of a maximal monotone multifunction that is not of type (ANA).

Theorem 38.2. *Let T: $D(T) \mapsto E^*$ be linear and maximal monotone. Then T is of type (FPV).*

Proof. Let U be a convex open subset of E such that $U \cap D(T) \neq \emptyset$. Suppose also that $(z, z^*) \in U \times E^*$ and the analog of (25.4.1) is satisfied, i.e.,

$$w \in D(T) \cap U \quad \Longrightarrow \quad \langle w - z, Tw - z^* \rangle \geq 0. \tag{38.2.1}$$

Our aim is to prove that

$$(z, z^*) \in G(T). \tag{38.2.2}$$

Since $U \cap D(T) \neq \emptyset$, we can fix $v \in U \cap D(T)$. Let $M := \langle v - z, Tv - z^* \rangle$. From (38.2.1), $M \geq 0$. Let $V := U - v$, and P be $M \times$ the Minkowski functional of V. Since V is a convex open set containing 0, P is a continuous positive sublinear functional on E. Explicitly, P: $E \mapsto \mathbb{R}$ is defined by

$$P(x) := \inf\{M\lambda\colon\ \lambda > 0,\ x \in \lambda V\}.$$

We first prove that

$$y \in D(T) \text{ and } u \in U \quad \Longrightarrow \quad \langle y - z, Ty - z^* \rangle + P(y - u) \geq 0. \tag{38.2.3}$$

Let $y \in D(T)$ and $u \in U$, and suppose that $\lambda > 0$ and $y - u \in \lambda V$. Write

$$w := \frac{y + \lambda v}{1 + \lambda} \in D(T).$$

Then

$$w = \frac{y + \lambda v}{1 + \lambda} \in \frac{u + \lambda V + \lambda v}{1 + \lambda} = \frac{u + \lambda U}{1 + \lambda} \subset U.$$

Thus, from (38.2.1), $\langle w - z, Tw - z^* \rangle \geq 0$. Since the function

$$x \mapsto \langle x - z, Tx - z^* \rangle$$

is convex,

$$\frac{\langle y - z, Ty - z^* \rangle + \lambda \langle v - z, Tv - z^* \rangle}{1 + \lambda} \geq 0$$

hence

$$\langle y - z, Ty - z^* \rangle + \lambda M \geq 0,$$

and (38.2.3) follows by taking the infimum over λ.

Now let $A := D(T) \times U$, and $B := \{x^* \in E^*\colon\ x^* \leq P \text{ on } E\}$ with the topology $w(E^*, E)$. From the extended Banach–Alaoglu theorem, Theorem 4.2, B is compact. Define h: $A \times B \mapsto \mathbb{R}$ by

$$h((y, u), x^*) := \langle y - z, Ty - z^* \rangle + \langle y - u, x^* \rangle.$$

From the one–dimensional Hahn–Banach theorem, Corollary 1.2, for all $(y, u) \in A$,

$$P(y - u) = \max\langle y - u, B \rangle$$

thus, from (38.2.3),

$$\inf_A \max_B h \geq 0.$$

The function h is convex on A, and affine and continuous on B. Thus from the minimax theorem, Theorem 3.1,

$$\max_B \inf_A h \geq 0,$$

hence there exists $x^* \in E^*$ such that

$$y \in D(T) \text{ and } u \in U \quad \Longrightarrow \quad \langle y - z, Ty - z^* \rangle + \langle y - u, x^* \rangle \geq 0,$$

or equivalently

$$y \in D(T) \text{ and } u \in U \quad \Longrightarrow \quad \langle y - z, Ty - z^* + x^* \rangle \geq \langle u - z, x^* \rangle. \quad (38.2.4)$$

Since T is maximal monotone, it follows by taking the infimum over $y \in D(T)$ and using Lemma 8.1(c) that

$$u \in U \quad \Longrightarrow \quad 0 \geq \langle u - z, x^* \rangle.$$

Now $z \in U$ and U is open, so it follows that $x^* = 0$. Substituting this back in (38.2.4), we obtain that

$$y \in D(T) \quad \Longrightarrow \quad \langle y - z, Ty - z^* \rangle \geq 0.$$

Since T is maximal monotone, it follows from this that $(z, z^*) \in G(T)$, i.e., (38.2.2) is satisfied. ▮

Theorem 38.3. *Let T: $D(T) \mapsto E^*$ be linear and maximal monotone of type (NI). Then T is of type (FP).*

Proof. Let U be a convex open subset of E^* such that $U \cap R(T) \neq \emptyset$. Suppose also that $(z, z^*) \in E \times U$ and the analog of (25.2.1) is satisfied, i.e.,

$$w \in D(T) \text{ and } Tw \in U \quad \Longrightarrow \quad \langle w - z, Tw - z^* \rangle \geq 0. \quad (38.3.1)$$

Our aim is to prove that

$$(z, z^*) \in G(T). \quad (38.3.2)$$

Since $U \cap R(T) \neq \emptyset$, we can fix $v \in D(T)$ so that $Tv \in U$. Let $M := \langle v - z, Tv - z^* \rangle$. From (38.3.1), $M \geq 0$. Let $V := U - Tv$, and P be $M \times$ the Minkowski functional of V. Since V is a convex open set containing 0, P is a continuous positive sublinear functional on E^*. Explicitly, P: $E^* \mapsto \mathbb{R}$ is defined by

$$P(x^*) := \inf\{M\lambda\colon\ \lambda > 0,\ x^* \in \lambda V\}.$$

We first prove that

$$y \in D(T) \text{ and } u^* \in U \quad \Longrightarrow \quad \langle y - z, Ty - z^* \rangle + P(Ty - u^*) \geq 0. \quad (38.3.3)$$

Let $y \in D(T)$ and $u^* \in U$, and suppose that $\lambda > 0$ and $Ty - u^* \in \lambda V$. Write

$$w := \frac{y + \lambda v}{1 + \lambda} \in D(T).$$

Then

$$Tw = \frac{Ty + \lambda Tv}{1 + \lambda} \in \frac{u^* + \lambda V + \lambda Tv}{1 + \lambda} = \frac{u^* + \lambda U}{1 + \lambda} \subset U.$$

Thus, from (38.3.1), $\langle w - z, Tw - z^* \rangle \geq 0$. Since the function

$$x \mapsto \langle x - z, Tx - z^* \rangle$$

is convex,

$$\frac{\langle y - z, Ty - z^* \rangle + \lambda \langle v - z, Tv - z^* \rangle}{1 + \lambda} \geq 0$$

hence

$$\langle y - z, Ty - z^* \rangle + \lambda M \geq 0,$$

and (38.3.3) follows by taking the infimum over λ.

Now let $A := D(T) \times U$, and $B := \{x^{**} \in E^{**} \colon x^{**} \leq P \text{ on } E^*\}$ with the topology $w(E^{**}, E^*)$. From the extended Banach–Alaoglu theorem, Theorem 4.2, B is compact. Define the function $h\colon A \times B \mapsto \mathbb{R}$ by

$$h((y, u^*), x^{**}) := \langle y - z, Ty - z^* \rangle + \langle Ty - u^*, x^{**} \rangle.$$

From the one–dimensional Hahn–Banach theorem, Corollary 1.2, for all $(y, u^*) \in A$,

$$P(Ty - u^*) = \max \langle Ty - u^*, B \rangle$$

thus, from (38.3.3),

$$\inf_A \max_B h \geq 0.$$

The function h is convex on A, and affine and continuous on B. Thus from the minimax theorem, Theorem 3.1,

$$\max_B \inf_A h \geq 0,$$

hence there exists $x^{**} \in E^{**}$ such that

$$y \in D(T) \text{ and } u^* \in U \implies [\langle y - z, Ty - z^* \rangle + \langle Ty - u^*, x^{**} \rangle] \geq 0,$$

or equivalently

$$y \in D(T) \text{ and } u^* \in U \implies \langle Ty - z^*, \widehat{y} - \widehat{z} + x^{**} \rangle \geq \langle u^* - z^*, x^{**} \rangle. \quad (38.3.4)$$

Since T is of type (NI), it follows by taking the infimum of the left hand side over $y \in D(T)$ that

$$u^* \in U \implies 0 \geq \langle u^* - z^*, x^{**} \rangle.$$

Now $z^* \in U$ and U is open, so it follows that $x^{**} = 0$. Substituting this back in (38.3.4), we obtain that

$$y \in D(T) \quad \Longrightarrow \quad \langle Ty - z^*, \widehat{y} - \widehat{z}\rangle \geq 0$$

that is to say,

$$y \in D(T) \quad \Longrightarrow \quad \langle y - z, Ty - z^*\rangle \geq 0.$$

Since T is maximal monotone, it follows from this that $(z, z^*) \in G(T)$, i.e., (38.3.2) is satisfied. ▌

Problem 38.4. Is the converse of Theorem 38.3 true? That is to say, if $T\colon\ D(T) \mapsto E^*$ is linear and of type (FP) then is T necessarily maximal monotone of type (NI)?

The last two results in this section appear in the paper [7] by Bauschke–Simons.

Theorem 38.5. *Let $D(T)$ be a subspace of E and $T\colon\ D(T) \mapsto E^*$ be linear and maximal monotone. Then T is strongly maximal monotone.*

Proof. Let C be a nonempty $w(E, E^*)$–compact subset of E, $w^* \in E^*$ and

$$\left.\begin{aligned}&\text{for all } y \in D(T),\\ &\qquad \text{there exists } w \in C \text{ such that } \quad \langle w - y, w^* - Ty\rangle \geq 0.\end{aligned}\right\} \tag{38.5.1}$$

Define $h\colon\ D(T) \times C \mapsto \mathbb{R}$ by

$$h(y, w) := \langle w - y, w^* - Ty\rangle = \langle w, w^*\rangle - \langle w, Ty\rangle - \langle y, w^*\rangle + \langle y, Ty\rangle.$$

From (38.5.1),

$$\inf_{D(T)} \max_{C} h \geq 0.$$

The function h is convex on $D(T)$ (exercise!), and also affine and $w(E, E^*)$–continuous on C. From the minimax theorem, Theorem 3.1,

$$\max_{C} \inf_{D(T)} h \geq 0,$$

that is to say,

$$\text{there exists } w \in C \text{ such that, } \quad \text{for all } y \in D(T), \quad \langle w - y, w^* - Ty\rangle \geq 0.$$

Since T is maximal monotone, it follows from this last inequality that $(w, w^*) \in G(T)$. Thus we have proved that:

$$\text{there exists } w \in C \text{ such that } \quad (w, w^*) \in G(T).$$

Similarly, let C be a nonempty $w(E^*, E)$–compact subset of E^*, $w \in E$ and

$$\left.\begin{array}{l}\text{for all } y \in D(T),\\ \qquad \text{there exists } w^* \in C \text{ such that } \quad \langle w-y, w^*-Ty\rangle \geq 0.\end{array}\right\} \tag{38.5.2}$$

Define h: $D(T) \times C \mapsto \mathbb{R}$ by

$$h(y,w^*) := \langle w-y, w^*-Ty\rangle = \langle w,w^*\rangle - \langle w,Ty\rangle - \langle y,w^*\rangle + \langle y,Ty\rangle.$$

From (38.5.2),

$$\inf_{D(T)} \max_{C} h \geq 0.$$

The function h is convex on $D(T)$, and also affine and $w(E^*,E)$–continuous on C. From the minimax theorem, Theorem 3.1,

$$\max_{C} \inf_{D(T)} h \geq 0,$$

that is to say,

$$\text{there exists } w^* \in C \text{ such that,} \quad \text{for all } y \in D(T), \quad \langle w-y, w^*-Ty\rangle \geq 0.$$

Since T is maximal monotone, it follows from this last inequality that $(w,w^*) \in G(T)$. Thus we have proved that:

$$\text{there exists } w^* \in C \text{ such that} \quad (w,w^*) \in G(T).$$

This completes the proof that T is strongly maximal monotone. ∎

Theorem 38.6. *Let T: $E \mapsto E^*$ be positive and linear. Then T is maximal monotone of type (ANA).*

Proof. Suppose that $(x,x^*) \in E \times E^* \setminus G(T)$. Then $Tx \neq x^*$. For all $n \geq 1$, we can find $z_n \in E$ such that $\|z_n\| = 1$ and

$$\langle z_n, Tx - x^*\rangle \to -\|Tx - x^*\| \quad \text{as } n \to \infty. \tag{38.6.1}$$

For all $n \geq 1$, let $w_n := x + z_n/n$. Then $\|Tw_n - Tx\| = \|Tz_n\|/n \leq \|T\|/n$ hence

$$\|Tw_n - Tx\| \to 0 \quad \text{and} \quad \|Tw_n - x^*\| \to \|Tx - x^*\| \neq 0 \quad \text{as } n \to \infty. \tag{38.6.2}$$

Now, for all sufficiently large $n \geq 1$, we have the inequality

$$\frac{|\langle w_n - x, Tw_n - Tx\rangle|}{\|w_n - x\|\|Tw_n - x^*\|} \leq \frac{\|Tw_n - Tx\|}{\|Tw_n - x^*\|}.$$

Combining this with (38.6.2), we obtain that

$$\frac{\langle w_n - x, Tw_n - Tx\rangle}{\|w_n - x\|\|Tw_n - x^*\|} \to 0 \quad \text{as } n \to \infty. \tag{38.6.3}$$

On the other hand, from (38.6.1) and (38.6.2),

$$\frac{\langle w_n - x, Tx - x^*\rangle}{\|w_n - x\|\|Tw_n - x^*\|} = \frac{\langle z_n, Tx - x^*\rangle}{\|Tw_n - x^*\|} \to \frac{-\|Tx - x^*\|}{\|Tx - x^*\|} = -1 \quad \text{as } n \to \infty.$$

Adding this to (38.6.3), we obtain that

$$\frac{\langle w_n - x, Tw_n - x^*\rangle}{\|w_n - x\|\|Tw_n - x^*\|} \to -1 \quad \text{as } n \to \infty.$$

This completes the proof that T is of type (ANA). ▌

Remark 38.7. As we have already observed in Problem 25.11, we do not know if T is necessarily of type (ANA) if $D(T)$ is a subspace of E and $T\colon D(T) \mapsto E^*$ is linear and maximal monotone.

IX. The sum problem for general Banach spaces

39. Introduction

We have already given in Theorem 37.1 an example of a situation in which we can assert that the sum of maximal monotone multifunctions on a general Banach space is maximal monotone. In this chapter, we describe three additional situations of this kind. These should be viewed in the following light: if we are trying to find a counterexample to the sum theorem in nonreflexive spaces (see the discussion in Section 26), then these results tell us where it is not worth looking!

40. Multifunctions with full domain

In this section we give a proof of Heisler's result that if $S_1\colon E \mapsto 2^{E^*}$ and $S_2\colon E \mapsto 2^{E^*}$ are maximal monotone and $D(S_1) = D(S_2) = E$ then, even if E is not reflexive, $S_1 + S_2$ is maximal monotone. Our proof is based on the characterization of maximal monotone multifunction with full domain contained in Theorem 40.2, and will give us the opportunity to introduce some of the classical techniques associated with maximal monotone multifunctions. However, our approach differs slightly from the usual one by the use of the scalar functions S_y defined below. See Phelps, [34], Section 7 for an approach based on upper semicontinuous multifunctions.

If $S\colon E \mapsto 2^{E^*}$ and $y \in E$, we define $S_y\colon D(S) \mapsto \mathbb{R} \cup \{\infty\}$ by

$$S_y(x) := \sup\langle y, Sx\rangle.$$

Lemma 40.1. *Let $S\colon E \mapsto 2^{E^*}$ be maximal monotone.*
(a) For all $x \in D(S)$, Sx is convex and $w(E^*, E)$–closed.
(b) For all $x \in \operatorname{int} D(S)$, Sx is convex and $w(E^*, E)$–compact.
(c) Suppose that (x_α, x_α^*) is a bounded net of elements of $G(S)$, $(x, x^*) \in E \times E^*$, $\|x_\alpha - x\| \to 0$ and $x_\alpha^* \to x^*$ in $w(E, E^*)$. Then $(x, x^*) \in G(S)$.
(d) Let $y \in E$. Then S_y is real–valued and upper semicontinuous on $\operatorname{int} D(S)$.

Proof. It is clear from Lemma 8.1(a,b) that

$$G(S) = \bigcap_{(s,s^*)\in G(S)} \{(x,x^*) \in E \times E^*\colon \langle x - s, x^* - s^*\rangle \geq 0\}; \tag{40.1.1}$$

consequently, for all $x \in D(S)$,

$$Sx = \bigcap_{(s,s^*)\in G(S)} \{x^* \in E^*\colon \langle x - s, x^* - s^*\rangle \geq 0\}. \tag{40.1.2}$$

(a) follows from (40.1.2), (b) follows from (a), the local boundendess theorem, Theorem 17.1, and the Banach-Alaoglu theorem, Theorem 4.1, and (c) follows from (40.1.1) (exercises!).

(d) It is clear from (b) that S_y is real–valued on $\operatorname{int} D(S)$. We will now prove that, for all $\lambda \in \mathbb{R}$,

$$\{x \in \operatorname{int} D(S)\colon\ S_y(x) \geq \lambda\} \quad \text{is closed relative to} \quad \operatorname{int} D(S). \tag{40.1.3}$$

So let

$$\lambda \in \mathbb{R} \quad \text{and} \quad v \in \operatorname{int} D(S) \cap \overline{\{x \in \operatorname{int} D(S)\colon\ S_y(x) \geq \lambda\}}. \tag{40.1.4}$$

From the local boundedness theorem, Theorem 17.1, there exist $\theta,\ Q > 0$ such that

$$s \in E \text{ and } \|s - v\| < \theta \quad \Longrightarrow \quad s \in \operatorname{int} D(S) \text{ and } \sup \|Ss\| \leq Q. \tag{40.1.5}$$

For all $n \geq 1$, it follows from (40.1.4) that there exists $x_n \in E$ such that

$$\|x_n - v\| < \frac{\theta}{n} \quad \text{and} \quad S_y(x_n) \geq \lambda. \tag{40.1.6}$$

Thus, using the definition of S_y, there exists $x_n^* \in Sx_n$ such that

$$\langle y, x_n^*\rangle > \lambda - \frac{1}{n}. \tag{40.1.7}$$

It is then clear from (40.1.5) and (40.1.6) that

$$\|x_n^*\| \leq Q. \tag{40.1.8}$$

It follows from this and the Banach–Alaoglu theorem, Theorem 4.1, that there exists a subnet (x_α, x_α^*) of the sequence $\{(x_n, x_n^*)\}_{n\geq 1}$ and $x^* \in E^*$ such that $x_\alpha^* \to x^*$ in $w(E^*, E)$. From (40.1.6), $\|x_\alpha - v\| \to 0$ and, using (40.1.8) again, the net (x_α, x_α^*) is bounded. Thus we derive from (c) that $(v, x^*) \in G(S)$. From (40.1.7), $\langle y, x^*\rangle \geq \lambda$ and so, from the definition of S_y,

$$S_y(v) \geq \langle y, x^*\rangle \geq \lambda.$$

This completes the proof of (40.1.3), and hence also the proof of (d). ∎

We now give a characterization of maximal monotone operators with full domain.

Theorem 40.2. *Let $S\colon E \mapsto 2^{E^*}$ be monotone and $D(S) = E$. Then S is maximal monotone if, and only, if*

$$\text{for all } x \in E, \quad Sx \text{ is convex and } w(E^*, E)\text{–compact} \tag{40.2.1}$$

and

$$\text{for all } y \in E, \quad S_y\colon E \mapsto \mathbb{R} \text{ is upper semicontinuous.} \tag{40.2.2}$$

Proof. ($\Longrightarrow$) This is immediate from Lemma 40.1(b,d).
($\Longleftarrow$) Suppose that (40.2.1) and (40.2.2) are satisfied. Let

$$(z, z^*) \in E \times E^* \quad \text{and} \quad \inf_{(s,s^*) \in G(S)} \langle s - z, s^* - z^* \rangle \geq 0. \tag{40.2.3}$$

Our aim is to prove that

$$(z, z^*) \in G(S). \tag{40.2.4}$$

Let y be an arbitrary element of E. Let $\lambda > 0$. Since $D(S) = E$, there exists $(s_\lambda, s_\lambda^*) \in G(S)$ such that $s_\lambda = z + \lambda y$. From (40.2.3),

$$\langle \lambda y, s_\lambda^* - z^* \rangle = \langle s_\lambda - z, s_\lambda^* - z^* \rangle \geq 0,$$

and consequently $\langle y, s_\lambda^* - z^* \rangle \geq 0$. Using the definition of S_y, we derive from this that

$$S_y(s_\lambda) \geq \langle y, z^* \rangle.$$

As $\lambda \to 0+$, $s_\lambda \to z$ and so, using (40.2.2),

$$S_y(z) \geq \langle y, z^* \rangle.$$

We have proved that,

$$\text{for all } y \in E, \quad \sup\langle y, Sz \rangle \geq \langle y, z^* \rangle.$$

Thus, from (40.2.1) and Theorem 4.8, $z^* \in Sz$. This establishes (40.2.4), and completes the proof of Theorem 40.2. ▌

Remark 40.3. It suffices for ($\Longleftarrow$) of Theorem 40.2 that

$$\text{for all } x \in E, \quad Sx \text{ be convex and } w(E^*, E)\text{–closed}$$

and

$$\text{for all } y \in E, \quad S_y \text{ be upper semicontinuous on every line–segment in } E.$$

We are now in a position to prove Heisler's result.

Theorem 40.4. *Let $S_1\colon E \mapsto 2^{E^*}$ and $S_2\colon E \mapsto 2^{E^*}$ be maximal monotone and $D(S_1) = D(S_2) = E$. Then $S_1 + S_2$ is maximal monotone.*

Proof. Write $S := S_1 + S_2$. The result is immediate from Theorem 40.2 since, for all $x \in E$, $Sx := S_1x + S_2x$, the sum of $w(E^*, E)$–compact convex sets is $w(E^*, E)$–compact and convex, for all $y \in E$, $S_y = (S_1)_y + (S_2)_y$, and the sum of upper semicontinuous functions is upper semicontinuous. ▮

41. Sums with normality maps

Theorem 41.1 is a multifunction version of the following result proved by Rockafellar in [43], Theorem 3, pp. 77 and 84: *Let C be a nonempty closed convex subset of E, $S\colon E \mapsto E^*$ be single–valued and monotone, $D(S) \supset C$, and S be continuous on all line segments in C with respect to the topology $w(E^*, E)$. Then $S + N_C$ is maximal monotone.* (We recall that the normality multifunction $N_C\colon E \mapsto 2^{E^*}$ is defined by

$$(x, x^*) \in G(N_C) \iff x \in C \text{ and } \langle x, x^* \rangle = \max_C x^*. \tag{8.1.1}$$

The proof of Theorem 41.1 was obtained by adapting the techniques of Section 40 to the proof in [43].

Theorem 41.1. *Let C be a nonempty closed convex subset of E, $S\colon E \mapsto 2^{E^*}$ be monotone and $D(S) \supset C$. Suppose that*

$$\text{for all } x \in C, \quad Sx \text{ is convex and } w(E^*, E)\text{–compact} \tag{41.1.1}$$

and

$$\left.\begin{array}{l}\text{for all } y \in C - C, \\ \qquad S_y \text{ is upper semicontinuous on all line–segments in } C.\end{array}\right\} \tag{41.1.2}$$

Then $S + N_C$ is maximal monotone.

Proof. Let

$$(z, z^*) \in E \times E^* \quad \text{and} \quad \inf_{(w,w^*) \in G(S+N_C)} \langle w - z, w^* - z^* \rangle \geq 0. \tag{41.1.3}$$

Our aim is to prove that

$$(z, z^*) \in G(S + N_C). \tag{41.1.4}$$

We first establish that

$$z \in C. \tag{41.1.5}$$

Let (x, x^*) be an arbitrary element of $G(N_C)$. Since $x \in C \subset D(S)$, there exists $s^* \in Sx$. Now let $\lambda \geq 0$. Then $(x, \lambda x^*) \in G(N_C)$, and so

$$(x, s^* + \lambda x^*) \in G(S + N_C).$$

Using (41.1.3), we now obtain that $\langle x - z, s^* + \lambda x^* - z^* \rangle \geq 0$, and consequently

$$\lambda \langle x - z, x^* - 0 \rangle + \langle x - z, s^* - z^* \rangle \geq 0.$$

Letting $\lambda \to \infty$,

$$\langle x - z, x^* - 0 \rangle \geq 0.$$

Since this holds for all $(x, x^*) \in G(N_C)$ and, as we have already observed in Section 8, N_C is maximal monotone, $(z, 0) \in G(N_C)$, which gives (41.1.5). (This is the trick used in the proof of Theorem 16.2.)

Now let u be an arbitrary element of C and $\lambda \in (0,1)$. Write $u_\lambda := \lambda u + (1 - \lambda)z \in C$. We next prove that

$$S_{u-z}(u_\lambda) \geq \langle u - z, z^* \rangle \tag{41.1.6}$$

Using (41.1.1), the set Su_λ is $w(E^*, E)$–compact, and so we can find

$$u_\lambda^* \in Su_\lambda \quad \text{such that} \quad \langle u - z, u_\lambda^* \rangle = S_{u-z}(u_\lambda). \tag{41.1.7}$$

Now $(u_\lambda, u_\lambda^*) = (u_\lambda, u_\lambda^* + 0) \in G(S + N_C)$ thus, using (41.1.3) again,

$$\langle \lambda u + (1 - \lambda)z - z, u_\lambda^* - z^* \rangle = \langle u_\lambda - z, u_\lambda^* - z^* \rangle \geq 0,$$

from which we derive that $\langle u - z, u_\lambda^* \rangle \geq \langle u - z, z^* \rangle$, and (41.1.6) now follows immediately from (41.1.7). As $\lambda \to 0$, $u_\lambda \to z$ and so, letting $\lambda \to 0$ in (41.1.6) and using (41.1.2),

$$S_{u-z}(z) \geq \langle u - z, z^* \rangle. \tag{41.1.8}$$

Define h: $C \times Sz \mapsto \mathbb{R}$ by

$$h(u, u^*) := \langle u - z, u^* - z^* \rangle.$$

Since (41.1.8) holds for any element u of C, we have proved that

$$\inf_C \max_{Sz} h \geq 0.$$

From (41.1.1), Sz is compact in the topology $w(E^*, E)$. Furthermore, the function h is affine on C, and affine and $w(E^*, E)$–continuous on Sz. Thus, from the minimax theorem, Theorem 3.1,

$$\max_{Sz} \inf_C h \geq 0,$$

that is to say, there exists $u^* \in Sz$ such that,

$$\text{for all } u \in C, \quad \langle u, z^* - u^* \rangle \leq \langle z, z^* - u^* \rangle.$$

Now this means that $z^* - u^* \in N_C(z)$, hence

$$z^* = u^* + (z^* - u^*) \subset (S + N_C)(z).$$

This completes the proof of (41.1.4), and hence that of Theorem 41.1. ▌

If we combine the result of Theorem 41.1 with that of Lemma 40.1, we obtain the following result:

Theorem 41.2. *Let C be a nonempty closed convex subset of E, $S\colon E \mapsto 2^{E^*}$ be maximal monotone and $\operatorname{int} D(S) \supset C$. Then $S + N_C$ is maximal monotone.*

We present in Lemma 41.3 a different situation in which we can assert that $S + N_C$ is maximal monotone. The condition $\operatorname{int} D(S) \supset C$ of Theorem 41.2 is weakened to $\operatorname{cen} D(S) \cap \operatorname{int} C \neq \emptyset$, where $v \in \operatorname{cen} D(S)$ means that, for all $z \in D(S)$, the segment $[v, z]$ is contained in $D(S)$. ("Cen" stands for "center".) The price that we pay is that we assume that S is *of type (FPV)*.

Lemma 41.3. *Let C be a closed convex subset of E, $S\colon E \mapsto 2^{E^*}$ be of type (FPV), and $\operatorname{cen} D(S) \cap \operatorname{int} C \neq \emptyset$. Then $S + N_C$ is maximal monotone.*

Proof. Let

$$(z, z^*) \in E \times E^* \quad \text{and} \quad \inf_{(w,w^*) \in G(S+N_C)} \langle w - z, w^* - z^* \rangle \geq 0. \tag{41.3.1}$$

Our aim is to prove that

$$(z, z^*) \in G(S + N_C). \tag{41.3.2}$$

If $(w, w^*) \in G(S)$ and $w \in \operatorname{int} C$ then

$$(w, w^*) = (w, w^* + 0) \in G(S + N_C),$$

hence $\langle w - z, w^* - z^* \rangle \geq 0$. Since S is of type (FPV) and $D(S) \cap \operatorname{int} C \neq \emptyset$, $(z, z^*) \in G(S)$. We will show that

$$z \in C. \tag{41.3.3}$$

Once this has been done, the relation $(z, z^*) = (z, z^* + 0)$ gives (41.3.2), and hence completes the proof of Lemma 41.3. So now let us prove (41.3.3). Fix $v \in \operatorname{cen} D(S) \cap \operatorname{int} C$, and write $V := \operatorname{int} C - v$. Then, from Kelly–Namioka, [28], 17.4, p. 155,

$$\overline{V} = \overline{\operatorname{int} C} - v = C - v.$$

Let P be the Minkowski functional of V. We note then (see, for instance, Phelps, [34], Example 1.1(d), p. 1–2) that

$$V = \{x \in E\colon\ P(x) < 1\}, \quad \overline{V} = \{x \in E\colon\ P(x) \leq 1\} \quad \text{and}$$
$$\operatorname{bd} V = \{x \in E\colon\ P(x) = 1\}.$$

If (41.3.3) fails then $z - v \notin \overline{V}$, hence $P(z - v) > 1$. Write

$$x := \Big(1 - \frac{1}{P(z - v)}\Big)v + \frac{1}{P(z - v)} z. \tag{41.3.4}$$

Clearly

$$P(x-v) = P\Big(\frac{1}{P(z-v)}(z-v)\Big) = 1,$$

and so $x - v \in \operatorname{bd} V$, from which $x \in v + \operatorname{bd} V = \operatorname{bd} C$. From the "support theorem", (see, for instance, Phelps, [34], p. 43 — which can be obtained by applying the one–dimensional form of the Hahn–Banach theorem, Corollary 1.2, to P) there exists $x^* \in N_C(x) \setminus \{0\}$. Since $v \in \operatorname{int} C$ and $x^* \neq 0$, we have

$$\langle v, x^* \rangle < \sup_C x^* = \langle x, x^* \rangle. \tag{41.3.5}$$

On the other hand, since $v \in \operatorname{cen} D(S)$ and $z \in D(S)$, it follows from (41.3.4) that $x \in D(S)$ and consequently there exists $s^* \in Sx$. Now let $\lambda \geq 0$. Then $(x, \lambda x^*) \in G(N_C)$, and so $(x, s^* + \lambda x^*) \in G(S + N_C)$. Using (41.3.1), we obtain that $\langle x - z, s^* + \lambda x^* - z^* \rangle \geq 0$, from which

$$\lambda \langle x - z, x^* \rangle + \langle x - z, s^* - z^* \rangle \geq 0.$$

Letting $\lambda \to \infty$, (again, the trick used in the proof of Theorem 16.2)

$$\langle x - z, x^* \rangle \geq 0. \tag{41.3.6}$$

We also derive from (41.3.4) that $z = P(z-v)x - (P(z-v) - 1)v$, and so

$$x - z = (P(z-v) - 1)(v - x).$$

Substituting this in (41.3.6) and noting that $P(z-v) - 1 > 0$,

$$\langle v - x, x^* \rangle \geq 0.$$

This contradiction of (41.3.5) completes the proof of (41.3.3), and hence that of Lemma 41.3. ▌

Problem 41.4. Lemma 41.3 is almost a converse to Theorem 26.1. This leads to the following problem: can we remove the "cen" from Lemma 41.3? In other words, is $S + N_C$ necessarily maximal monotone if C is a closed convex subset of E, $S\colon E \mapsto 2^{E^*}$ is of type (FPV), and $D(S) \cap \operatorname{int} C \neq \emptyset$?

If $D(S)$ is convex then $\operatorname{cen} D(S) = D(S)$, and so Theorem 41.5 is immediate from Lemma 41.3. (We note from Theorem 26.3 that $\overline{D(S)}$ is automatically convex.)

Theorem 41.5. *Let C be a closed convex subset of E, $S\colon E \mapsto 2^{E^*}$ be of type (FPV), $D(S)$ be convex and $D(S) \cap \operatorname{int} C \neq \emptyset$. Then $S + N_C$ is maximal monotone.*

If we combine Theorem 41.5 with Theorem 38.2, we obtain the following result.

Theorem 41.6. *Let C be a closed convex subset of E, $D(S)$ be a subspace of E, S: $D(S) \mapsto E^*$ be linear and maximal monotone and $D(S) \cap \operatorname{int} C \neq \emptyset$. Then $S + N_C$ is maximal monotone.*

Remark 41.7. It is worth pointing out that the multifunction $S + N_C$ appears in a different context in Lemma 16.1.

42. Sums with linear maps

The two results in this section are due to Heinz Bauschke (personal communication). In Section 8, we defined a *skew linear operator* to be a linear operator T: $E \mapsto E^*$ such that

$$x \in E \quad \Longrightarrow \quad \langle x, Tx \rangle = 0.$$

Theorem 42.1. *Let S: $E \mapsto 2^{E^*}$ be maximal monotone and T: $E \mapsto E^*$ be skew and linear. Then $S + T$ is maximal monotone*

Proof. Let

$$(z, z^*) \in E \times E^* \quad \text{and} \quad \inf_{(w,w^*) \in G(S+T)} \langle w - z, w^* - z^* \rangle \geq 0. \tag{42.1.1}$$

Our aim is to prove that

$$(z, z^*) \in G(S + T). \tag{42.1.2}$$

Let (x, x^*) be an arbitrary element of $G(S)$. Then $(x, x^* + Tx) \in G(S + T)$ and so, from (42.1.1),

$$\langle x - z, Tx \rangle + \langle x - z, x^* - z^* \rangle = \langle x - z, x^* + Tx - z^* \rangle \geq 0. \tag{42.1.3}$$

However, $\langle x - z, Tx \rangle - \langle x - z, Tz \rangle = \langle x - z, T(x - z) \rangle = 0$, and so

$$\langle x - z, Tx \rangle = \langle x - z, Tz \rangle.$$

Substituting this in (42.1.3), $\langle x - z, Tz \rangle + \langle x - z, x^* - z^* \rangle \geq 0$, that is to say,

$$\langle x - z, x^* + Tz - z^* \rangle \geq 0.$$

Since this holds for all $(x, x^*) \in G(S)$ and S is maximal monotone, it follows that $(z, z^* - Tz) \in G(S)$, and so $z^* = z^* - Tz + Tz \in (S + T)(z)$. This completes the proof of (42.1.2), and hence that of Theorem 42.1. ∎

Theorem 42.2. *Let $f\colon E \mapsto \mathbb{R} \cup \{\infty\}$ be convex and lower semicontinuous with* $\operatorname{dom} f \neq \emptyset$, *and $T\colon E \mapsto E^*$ be positive and linear. Then $\partial f + T$ is maximal monotone.*

Proof. We define the convex continuous function $e\colon E \mapsto \mathbb{R}$ by

$$e(x) := \frac{1}{2}\langle x, Tx\rangle \quad (x \in E),$$

and the positive linear operator $P\colon E \mapsto E^*$ by

$$\langle y, Px\rangle := \frac{1}{2}\big[\langle y, Tx\rangle + \langle x, Ty\rangle\big] \quad (x,\ y \in E).$$

Then, for all $x \in E$, $\partial e(x) = \{Px\}$ (exercise!), and the linear operator $T - P\colon E \mapsto E^*$ is skew (exercise). Using Theorem 28.2, we then have

$$\partial f + T = \partial f + P + (T - P) = \partial f + \partial e + (T - P) = \partial(f + e) + (T - P).$$

However, from Rockafellar's maximal monotonicity theorem (see Section 29), $\partial(f + e)$ is maximal monotone, and the result follows from Theorem 42.1. ∎

X. Open problems

Problem 10.10. Find the smallest constant C such that if E is a reflexive Banach space and M is a maximal monotone subset of $E \times E^*$ then, for all $(x, x^*) \in M$ such that

$$\|x\|^2 + \|x^*\|^2 + 2\langle x, x^*\rangle = 0,$$

we must have

$$(s, s^*) \in M \quad \Longrightarrow \quad \|x\|^2 + \|x^*\|^2 \le C\big(\|s\|^2 + \|s^*\|^2\big).$$

(We have from the proof of Lemma 10.1 that $C \le (1 + \sqrt{2})^2$.)

Problem 13.6. Let g_1, g_2: $E \mapsto \mathrm{I\!R} \cup \{\infty\}$ be convex Borel functions and $\mathrm{dom}\, g_1 - \mathrm{dom}\, g_2$ surround 0. Does there necessarily exist $n \ge 1$ such that

$$\{E|\ g_1 \le n,\ \|\ \| \le n\} - \{E|\ g_2 \le n,\ \|\ \| \le n\} \quad \text{is a neighborhood of } 0?$$

In particular: Let B_1 and B_2 be convex Borel sets in E and $B_1 - B_2$ be absorbing. Is $B_1 - B_2$ necessarily a neighborhood of 0?

Problem 15.5. Find a Banach space E and a maximal monotone multifunction S: $E \mapsto 2^{E^*}$ such that

$$\mathrm{dom}\, \chi_S \quad \text{is a proper subset of} \quad \mathrm{dom}\, \psi_S.$$

Problem 16.3. Let S: $E \mapsto 2^{E^*}$ be maximal monotone. Then is it necessarily true that

$$\overline{\mathrm{dom}\, \chi_S} = \overline{\mathrm{co} D(S)}?$$

Problem 18.9. Is $\overline{D(S)}$ necessarily convex when E is not reflexive, S is maximal monotone and $\mathrm{sur}\,(\mathrm{dom}\, \chi_S) = \emptyset$?

Problem 18.10. Let S: $E \mapsto 2^{E^*}$ be maximal monotone. If E is not reflexive, do we always have:

$$\mathrm{int}\,(\mathrm{dom}\, \psi_S) \subset D(S)?$$

Problem 19.6. Let H be a Hilbert space and $S: H \mapsto 2^H$ and $T: H \mapsto 2^H$ be monotone. Suppose that $R(S) + R(T) \ni 0$, and also that there exists $M \geq 0$ such that, for arbitrarily small $\lambda > 0$, there exists $(u_\lambda, u_\lambda^*) \in G(S)$ such that $\|u_\lambda^*\| \leq M$ and $(u_\lambda, -u_\lambda^* - \lambda u_\lambda) \in G(T)$. Then does there necessarily exist $\beta \geq 0$ such that

$$(w, w^*) \in G(S+T) \quad \Longrightarrow \quad \langle w, w^* \rangle + \beta(1 + \|w^*\|) \geq 0?$$

Problem 21.5. Let E be reflexive, S_1: $E \mapsto 2^{E^*}$ and S_2: $E \mapsto 2^{E^*}$ be maximal monotone and

$$\operatorname{dom} \psi_{S_1} - \operatorname{dom} \psi_{S_2} \quad \text{be absorbing.}$$

Then is $S_1 + S_2$ necessarily maximal monotone? Of course, this problem only makes sense after we have found solutions to Problem 15.5.

Problem 25.7. If S is maximal monotone of type (NI) then does it necessarily follow that S is maximal monotone of type (D)?

Problem 25.9. Is every maximal monotone multifunction strongly maximal monotone?

Problem 25.11. Is every maximal monotone multifunction of type (ANA)? (We do not even know what the situation is for discontinuous positive linear operators.)

Problem 27.7. If S is maximal monotone of type (NI), is $\overline{R(S)}$ necessarily convex?

Problem 34.9. Let $f \in \mathcal{PCLSC}(E)$, and $(t^*, t^{**}) \in G(\partial f^*)$. Does there necessarily exist a net $\{(s_\alpha, s_\alpha^*)\}$ of elements of $G(\partial f)$ such that $\widehat{s_\alpha} \to t^{**}$ in $\mathcal{T}_{CC}(E^{**})$, $s_\alpha^* \to t^*$ in norm and $f(s_\alpha) \to f^{**}(t^{**})$? In other words, can we replace "$\mathcal{T}_{C\mathcal{L}\mathcal{B}}(E^{**})$" by "$\mathcal{T}_{CC}(E^{**})$" in the statement of Theorem 34.7?

Problem 37.2. If S and T are maximal monotone and linear and

$$D(S) - D(T) \quad \text{is absorbing}$$

(or, equivalently, $D(S) - D(T) = E$) then is $S + T$ necessarily maximal monotone?

Problem 38.1. If T: $D(T) \mapsto E^*$ is linear and maximal monotone of type (NI), does it automatically follow that T is of type (DS)?

Problem 38.4. If T: $D(T) \mapsto E^*$ is linear and of type (FP) then is T necessarily maximal monotone of type (NI)?

Problem 41.4. Is $S + N_C$ necessarily maximal monotone if C is a closed convex subset of E, S: $E \mapsto 2^{E^*}$ is of type (FPV), and $D(S) \cap \operatorname{int} C \neq \emptyset$?

References

[1] H. Attouch and H. Brézis, *Duality for the sum of convex funtions in general Banach spaces.*, Aspects of Mathematics and its Applications, J. A. Barroso, ed., Elsevier Science Publishers (1986), 125–133.

[2] H. Attouch, H. Riahi and M. Théra, *Somme ponctuelle d'opérateurs maximaux monotones*, Serdica Math. Journal, **22** (1996), 267–292.

[3] J.-P. Aubin and I. Ekeland, *Applied Nonlinear Analysis*, Wiley, New York – Chichester – Brisbane – Toronto – Singapore (1984).

[4] J.-P. Aubin and H. Frankowska, *Set–Valued Analysis*, Birkhäuser, Boston – Basel – Berlin (1990).

[5] H. H. Bauschke and J. M. Borwein, *Continuous linear monotone operators on Banach spaces*, preprint.

[6] H. H. Bauschke and J. M. Borwein, *Maximal monotonicity of dense type, local maximal monotonicity, and monotonicity of the conjugate are all the same for continuous linear operators*, preprint.

[7] H. H. Bauschke and S. Simons, *Linear maximal monotone operators are strongly maximal monotone*, in preparation.

[8] G. Beer, *The slice topology: A viable alternative to Mosco convergence in nonreflexive spaces*, Nonlinear Analysis, **19** (1992), 271–290.

[9] J. M. Borwein, *A Lagrange multiplier theorem and a sandwich theorem for convex relations*, Math. Scand., **48** (1981), 198–204.

[10] J. M. Borwein and S. Fitzpatrick, *Local boundedness of monotone operators under minimal hypotheses*, Bull. Australian Math. Soc. **39** (1988), 439–441.

[11] J. Borwein, S. Fitzpatrick and J. Vanderwerff, *Examples of convex functions and classification of normed spaces*, J. Convex Analysis **1**(1994), 61–73.

[12] H. Brézis, M. G. Crandall and A. Pazy, *Perturbations of nonlinear maximal monotone sets in Banach spaces*, Comm. Pur. Appl. Math. **23** (1970), 123–144.

[13] H. Brézis and A. Haraux, *Image d'une somme d'opérateurs monotone at applications*, Israel J. Math. **23**(1976), 165–186.

[14] A. Brøndsted and R.T. Rockafellar, *On the Subdifferentiability of Convex Functions*, Proc. Amer. Math. Soc. **16**(1965), 605–611.

[15] F. E. Browder, *Nonlinear maximal monotone operators in Banach spaces*, Math. Annalen **175** (1968), 89–113.

[16] L.-J. Chu, *On the sum of monotone operators*, Michigan Math. J., **43**(1996), 273–289.

[17] M. Coodey, *Examining maximal monotone operators using pictures and convex functions*, Ph. D. dissertation, University of California, Santa Barbara, June 1997.

[18] M. Coodey and S. Simons, *The convex function determined by a multifunction*, Bull. Austral. Math. Soc. **54** (1996), 87–97.

[19] K. Deimling, *Nonlinear Functional Analysis*, Springer–Verlag, New York – Heidelberg – Berlin – Tokyo (1985).
[20] I. Ekeland, *Nonconvex minimization problems*, Bull. Amer. Math. Soc. **1** (1979), 443–474.
[21] K. Fan, *Minimax theorems*, Proc. Nat. Acad. Sci. U.S.A. **39** (1953), 42–47.
[22] K. Fan, I. Glicksberg and A. J. Hoffman, *Systems of inequalities involving convex functions*, Proc. Amer. Math. Soc. **8** (1957), 617–622.
[23] S. P. Fitzpatrick and R. R. Phelps, *Bounded approximants to monotone operators on Banach spaces*, Ann. Inst. Henri Poincaré, Analyse non linéaire **9** (1992), 573–595.
[24] S. P. Fitzpatrick and R. R. Phelps, *Some properties of maximal monotone operators on nonreflexive Banach spaces*, Set–Valued Analysis **3**(1995), 51–69.
[25] S. P. Fitzpatrick and S. Simons, *On the maximum of two convex functions*, in preparation.
[26] J.- P. Gossez, *Opérateurs monotones non linéaires dans les espaces de Banach non réflexifs* J. Math. Anal. Appl. **34** (1971), 371–395.
[27] R. B. Holmes, *Geometric functional analysis and its applications*, Graduate Texts in Mathematics, **24** (1975) Springer–Verlag, New York – Heidelberg.
[28] J. L. Kelley, I. Namioka, and co-authors, *Linear Topological Spaces*, D. Van Nostrand Co., Inc., Princeton – Toronto – London – Melbourne (1963).
[29] H. König, *Über das Von Neumannsche Minimax-Theorem*, Arch. Math. **19** (1968), 482–487.
[30] H. König, *On certain applications of the Hahn–Banach and minimax theorems*, Arch. Math. **21** (1970), 583–591.
[31] H. König, *Sublineare Funktionale*, Arch. Math. **23** (1972), 500–508.
[32] S. Kum, *Maximal monotone operators in the one–dimensional case*, J. Korean Math. Soc. **34** (1997), 371–381.
[33] D. T. Luc, *A resolution of Simons' maximal monotonicity problem*, J. Convex Analysis **3** (1996), 367–370.
[34] R. R. Phelps, *Convex Functions, Monotone Operators and Differentiability*, Lecture Notes in Mathematics **1364** (1993), Springer–Verlag (Second Edition).
[35] R. R. Phelps, *Lectures on Maximal Monotone Operators*, 2nd Summer School on Banach Spaces, Related Areas and Applications, Prague and Paseky, August 15–28, 1993. (Preprint, 30 pages.) TeX file: Banach space bulletin board archive:
<ftp://ftp:@math.okstate.edu/pub/banach/phelpsmaxmonop.tex>.
Posted Nov. 1993.
[36] R. R. Phelps and S. Simons, *Unbounded linear monotone operators on nonreflexive Banach spaces*, in preparation.
[37] J. D. Pryce, *Weak compactness in locally convex spaces*, Proc. Amer. Math. Soc. **17** (1966), 148–155.
[38] S. Reich, *The range of sums of accretive and monotone operators*, J. Math. Anal. Appl. **68** (1979), 310–317.
[39] S. M. Robinson, *Regularity and stability for convex multivalued functions*, Math. Oper. Res. **1** (1976), 130–143.
[40] R. T. Rockafellar, *Extension of Fenchel's duality theorem for convex functions*, Duke Math. J. **33** (1966), 81–89.
[41] R. T. Rockafellar, *Local boundedness of Nonlinear, Monotone Operators*, Michigan Math. J. **16** (1969), 397–407.
[42] R. T. Rockafellar, *On the Virtual Convexity of the Domain and Range of a Nonlinear Maximal Monotone Operator*, Math. Ann. **185** (1970), 81–90.

[43] R. T. Rockafellar, *On the Maximality of Sums of Nonlinear Monotone Operators*, Trans. Amer. Math. Soc. **149** (1970), 75–88.
[44] R. T. Rockafellar, *On the maximal monotonicity of subdifferential mappings*, Pac. J. Math. **33**(1970), 209-216.
[45] R. T. Rockafellar, *Conjugate duality and optimization*, Conference Board of the Mathematical Sciences **16**(1974), SIAM publications.
[46] W. Rudin, *Functional analysis*, McGraw-Hill, New York (1973).
[47] S. Simons, *Formes souslinéaires minimales*, Seminaire Choquet 1970/1971, no. 23, 8 pages.
[48] S. Simons, *Critères de faible compacité en termes du théorème de minimax*, Seminaire Choquet 1970/1971, no. 24, 5 pages.
[49] S. Simons, *Subdifferentials are locally maximal monotone*, Bull. Australian Math. Soc. **47** (1993), 465–471.
[50] S. Simons, *Subtangents with controlled slope*, Nonlinear Analysis, **22**(1994), 1373–1389.
[51] S. Simons, *Swimming below icebergs*, Set–Valued Analysis **2** (1994), 327–337.
[52] S. Simons, *Minimax theorems and their proofs*, Minimax and applications, Ding-Zhu Du and Panos M. Pardalos eds., Kluwer Academic Publishers, Dordrecht – Boston (1995), 1–23.
[53] S. Simons, *The range of a monotone operator*, J. Math. Anal. Appl. **199** (1996), 176–201.
[54] S. Simons, *Pictures of monotone operators*, Set–Valued Analysis **4** (1996), 271–282.
[55] S. Simons, *Subdifferentials of convex functions*, in "Recent Developments in Optimization Theory and Nonlinear Analysis", Y. Censor and S. Reich eds, American Mathematical Society, Providence, Rhode Island, Contemporary Mathematics **204** (1997), 217–246.
[56] S. Simons, *Sum theorems for monotone operators and convex functions*, Trans. Amer. Math. Soc., in press.
[57] S. Simons, *Pairs of monotone operators*, Trans. Amer. Math. Soc., in press.
[58] C. Ursescu, *Multifunctions with convex closed graph* Czechoslovak Math. J. **25** (1975), 438–441.
[59] A. and M. E. Verona, *Remarks on subgradients and ε-subgradients*, Set–Valued Analysis **1**(1993), 261–272.
[60] A. and M. E. Verona, *Regular maximal monotone operators*, Preprint.
[61] D. Zagrodny, *The maximal monotonicity of the subdifferentials of convex functions: Simons' problem*, Set–Valued Analysis **4**(1996), 301–314.
[62] E. Zeidler, *Nonlinear Functional Analysis and its Applications*, Vol II/B Nonlinear Monotone Operators, Springer–Verlag, New York–Berlin–Heidelberg (1990).

Subject Index

Symbol index